ENVIRONMENTAL SCIENCE, ENGINEERING AND TECHNOLOGY

CLIMATE CHANGE PROGRAMS AND ACTIVITIES

ANALYSES OF FEDERAL FUNDING

Environmental Science, Engineering and Technology

Additional books in this series can be found on Nova's website under the Series tab.

Additional E-books in this series can be found on Nova's website under the E-book tab.

Climate Change and its Causes, Effects and Prediction

Additional books in this series can be found on Nova's website under the Series tab.

Additional E-books in this series can be found on Nova's website under the E-book tab.

ENVIRONMENTAL SCIENCE, ENGINEERING AND TECHNOLOGY

CLIMATE CHANGE PROGRAMS AND ACTIVITIES

ANALYSES OF FEDERAL FUNDING

ANDRE H. ANDERSON
AND
DERRICK J. THOMAS
EDITORS

Nova Science Publishers, Inc.
New York

For permission to use material from this book please contact us:
Telephone 631-231-7269; Fax 631-231-8175
Web Site: http://www.novapublishers.com

LIBRARY OF CONGRESS CATALOGING-IN-PUBLICATION DATA

ISBN: 978-1-61942-760-0

Published by Nova Science Publishers, Inc. † New York

CONTENTS

PREFACE

Climate change is a complex, crosscutting issue that poses risks to many existing environmental and economic systems, including agriculture, infrastructure, ecosystems, and human health. A 2009 assessment by the United States Global Change Research Program found that climate-related changes, such as rising temperature and sea level, will combine with pollution, population growth, urbanization, and other social, economic and environmental stresses to create larger impacts than from any of these factors alone. Funding for climate change activities is spread across the federal government. This book provides an analysis of federal funding for climate change activities and how these activities are organized.

In: Climate Change
Eds: A.H. Anderson and D. J. Thomas
ISBN: 978-1-61942-760-0

Chapter 1

CLIMATE CHANGE: IMPROVEMENTS NEEDED TO CLARIFY NATIONAL PRIORITIES AND BETTER ALIGN THEM WITH FEDERAL FUNDING DECISIONS*

United States Government Accountability Office

WHY GAO DID THIS STUDY

Climate change poses risks to many environmental and economic systems, including agriculture, infrastructure, and ecosystems. Federal law has periodically required the Office of Management and Budget (OMB) to report on federal climate change funding.

GAO was asked to examine (1) federal funding for climate change activities and how these activities are organized; (2) the extent to which methods for defining and reporting climate change funding are interpreted consistently across the federal government; (3) federal climate change strategic priorities, and the extent to which funding is aligned with these priorities; and (4) what options, if any, are available to better align federal climate change funding with strategic priorities. GAO analyzed OMB funding

* This is an edited, reformatted and augmented version of the Highlights of GAO-11-317, a report to the Ranking Member, Committee on Natural Resources, House of Representatives, dated May 2011.

reports and responses to a Web-based questionnaire sent to federal officials, reviewed available literature, and interviewed stakeholders.

WHAT GAO RECOMMENDS

Among GAO's recommendations are that the appropriate entities within the Executive Office of the President (EOP), in consultation with Congress, clearly establish federal strategic climate change priorities and assess the effectiveness of current practices for defining and reporting related funding. Relevant EOP entities did not provide official written comments, but instead provided technical comments, which GAO incorporated as appropriate.

WHAT GAO FOUND

Funding for climate change activities reported by OMB increased from $4.6 billion in 2003 to $8.8 billion in 2010, and is organized in a complex, crosscutting system. OMB reports funding in four categories: technology to reduce emissions, science to better understand climate change, international assistance for developing countries, and wildlife adaptation to respond to actual or expected changes. Over this period, technology funding, the largest category, increased from $2.56 billion to $5.5 billion and increased as a share of total funding. OMB also reported $26.1 billion as funding for climate change programs and activities in the American Recovery and Reinvestment Act of 2009, and tax expenditures to encourage emissions reductions, with $7.2 billion in federal revenue losses in 2010. Many federal entities manage related activities, including interagency programs that coordinate agency actions.

Questionnaire responses suggest that methods for defining and reporting climate change funding are not interpreted consistently across the federal government. Respondents identified three methods for defining and reporting climate change funding, foremost of which is guidance contained in OMB Circular A-11. While most said their own organization consistently applied these methods internally, far fewer said that they were applied consistently across the government. Some, for example, noted that other agencies use their own interpretation of definitions, resulting in inconsistent accounting across the government, because of several factors, such as the difficulty in distinguishing between programs related and unrelated to climate change.

Respondents, literature, and stakeholders identified two key factors that complicate efforts to align funding with priorities. First, notwithstanding existing coordinating mechanisms, questionnaire results indicated that federal officials do not have a shared understanding of strategic priorities. This is in part due to inconsistent messages articulated in strategic plans and other policy documents. A 2008 Congressional Research Service analysis had similarly found no "overarching policy goal for climate change that guides the programs funded or the priorities among programs." Second, respondents indicated that since mechanisms for aligning funding with priorities are nonbinding, they are limited when in conflict with agencies' own priorities.

Questionnaire respondents also identified options to better align funding with strategic priorities. Such options included (1) a governmentwide strategic planning process that promotes a shared understanding among agencies of strategic priorities by articulating what they are expected to do within the overall federal response to climate change and (2) an integrated budget review process that better aligns these priorities with funding decisions through a more consistent method of reporting and reviewing climate change funding. Federal entities are beginning to implement some of these options. However, without further improvement in how federal climate change funding is defined and reported, strategic priorities are set, and funding is aligned with priorities, it will be difficult for the public and Congress to fully understand how climate change funds are accounted for and how they are spent.

ABBREVIATIONS

CCTP	Climate Change Technology Program
CEQ	Council on Environmental Quality
CENRS	Committee on Environment, Natural Resources, and Sustainability
CRS	Congressional Research Service
DOE	U.S. Department of Energy
EOP	Executive Office of the President
EPA	Environmental Protection Agency

MAX	OMB MAX Information System
NAPA	National Academy of Public Administration
NEPA	National Environmental Policy Act of 1969
NOAA	National Oceanic and Atmospheric Administration
NRC	National Research Council
NSTC	National Science and Technology Council
OECC	Office of Energy and Climate Change Policy
OMB	Office of Management and Budget
OSTP	Office of Science and Technology Policy
OTA	Office of Technology Assessment
Recovery Act	American Recovery and Reinvestment Act of 2009
SGCR	Subcommittee on Global Change Research
USGCRP	United States Global Change Research Program

May 20, 2011

The Honorable Edward J. Markey
Ranking Member
Committee on Natural Resources
House of Representatives

Dear Mr. Markey:

Climate change is a complex, crosscutting issue that poses risks to many existing environmental and economic systems, including agriculture, infrastructure, ecosystems, and human health. A 2009 assessment by the United States Global Change Research Program (USGCRP)[1] found that climate-related changes—such as rising temperature and sea level—will combine with pollution, population growth, urbanization, and other social, economic, and environmental stresses to create larger impacts than from any of these factors alone.[2] Funding for climate change activities is spread across the federal government. According to the Office of Management and Budget's (OMB) June 2010 *Federal Climate Change Expenditures Report to Congress*, 9 of the 15 cabinet-level executive departments, along with 7 other federal agencies, received funding for climate change activities in fiscal year 2010.[3] In addition, entities within the Executive Office of the President (EOP) such as the Office of Science and Technology Policy (OSTP) and federal interagency coordinating bodies like USGCRP work together to ensure federal climate change activities are guided by the latest climate science.

Several recent reports have found shortcomings in federal efforts to account for and organize climate change programs and activities. For example, in 2008 the Congressional Research Service (CRS) reported that the packaging of mostly existing programs into a federal climate change strategy has resulted in a lack of a unifying mission across the federal government.[4] This finding is reinforced by two of our recent reports—a 2010 report on environmental satellites used for measuring variations in climate over time and a 2009 report on climate change adaptation—which showed that certain federal climate-related activities were not well coordinated across the government.[5] We have found in the past that when agencies do not collaborate well when addressing a complicated, interdisciplinary issue like climate change, they may carry out programs in a fragmented, uncoordinated way, resulting in a patchwork of programs that can limit the overall effectiveness of the federal effort.[6]

In this context, you asked us to review federal climate change activities. Our objectives were to examine (1) federal funding for climate change activities and how these activities are organized; (2) the extent to which methods for defining and reporting climate change funding are interpreted consistently across the federal government; (3) federal strategic climate change priorities, and the extent to which funding is aligned with these priorities; and (4) what options, if any, are available to better align federal climate change funding with strategic priorities.

To address these objectives, we analyzed climate change funding data and tax expenditures presented in OMB's reports to Congress on federal climate change expenditures, reviewed other relevant reports, and interviewed stakeholders knowledgeable about federal funding for climate change programs and activities. To identify relevant reports and stakeholders, we reviewed our prior climate change work and conducted a literature search and review. Using this information, we developed a Web- based questionnaire to gather information and opinions of key federal officials involved in defining and reporting climate change funding, developing strategic priorities, or aligning funding with strategic priorities.

We worked with federal officials within EOP, interagency coordinating programs, and individual agencies to pretest the questionnaire; the final list of questionnaire recipients included 106 federal officials. Of these officials, 73 responded to the questionnaire, for a response rate of about 69 percent.[7] Not all officials responded to all questions. Given our methodology, we may not have identified every person who has knowledge of or experience with the topics we covered. However, we believe we were able to reach many of the relevant federal officials because we worked with EOP and interagency coordinating bodies to identify the federal officials to whom we sent the questionnaire. We did not conduct statistical analyses because the sample of respondents was not a representative sample. We analyzed questionnaire responses to group similar responses together into overall themes, and used specific responses as illustrative examples throughout the report.

The term "funding" in this report refers to budget authority, or the authority provided by federal law to enter into financial obligations that will result in outlays involving federal government funds, as reported by OMB in its reports. We use the term "account" to describe the budget accounts, line items, programs, and activities presented in OMB reports. Unless otherwise stated, we report funding in nominal terms (not adjusted for inflation), and all years refer to fiscal years. Unless otherwise specified, figures represent actual funding (not estimates), with the exception of 1993, 1994, and 2010, where we present estimated funding because actual data are not available. "Respondents" in this report refers to federal officials who completed the Web-based questionnaire. "Stakeholders" refers to other individuals we interviewed who have experience with federal funding for climate change programs and activities. Appendix I provides a more detailed description of our scope and methodology.

We conducted this performance audit from November 2009 to May 2011 in accordance with generally accepted government auditing standards. Those standards require that we plan and perform the audit to obtain sufficient, appropriate evidence to provide a reasonable basis for our findings and conclusions based on our audit objectives. We believe that the evidence obtained provides a reasonable basis for our findings and conclusions based on our audit objectives.

BACKGROUND

In August 2005, we issued a report on federal climate change funding for 1993 through 2004, as reported by OMB.[8] Specifically, we reported on how (1) total funding and funding by category changed and whether funding data were comparable over time and (2) funding by agency changed and whether funding data were comparable over time. We found, among other things, that it was unclear whether funding changed as much as OMB reported because modifications in the format and content of OMB reports limited the comparability of funding data over time. For example, OMB reported that it expanded the definitions of some accounts to include more activities, but did not specify how it changed the definitions. We were also unable to compare climate-related tax expenditures over time because OMB reported data on *proposed*, but not on *existing* tax expenditures.[9]

Based on these findings, we recommended that OMB (1) use the same format for presenting data from year-to-year, to the extent that it could do so and remain in compliance with reporting requirements; (2) explain changes in report content or format when they are introduced; (3) include information on existing climate-related tax expenditures in its reports; and (4) use the same criteria for determining which tax expenditures to include as it uses for determining which accounts to include. Presenting tax expenditures alongside the related spending programs is a first step in providing a useful and accurate picture of the extent of federal support for climate change.

In its April 2006 *Federal Climate Change Expenditures Report to Congress*—the first following our August 2005 report—OMB responded to our recommendations about report consistency and tax expenditures.[10] The report stated that "to address GAO's recommendations, reporting changes have been noted in table footnotes throughout this report and a summary table of climate funding from 2003 through 2007 has been provided." The report also included existing tax expenditures that could contribute to reducing greenhouse gas emissions. OMB's most recent reports generally have kept the same structures, categories, definitions, and format as in past years and more clearly label funding data.[11]

Federal Funding for Climate Change Activities Increased Substantially from 2003 through 2010, and Reflects a Complex, Crosscutting System

Funding for climate change activities reported by OMB increased from $4.6 billion in 2003 to $8.8 billion in 2010. In recent years, both funding provided in the American Recovery and Reinvestment Act of 2009 (Recovery Act) and energy-related tax expenditures (some established in the Recovery Act) contributed significantly to the overall level of federal resources focused on climate change.[12] Many federal entities manage climate change programs and activities, including those within EOP, and interagency committees and programs that coordinate the actions of individual agencies, reflecting a complex, crosscutting system.

Funding Increased Significantly from 2003 through 2010, as Reported by OMB

From 2003 through 2010, total federal funding for climate change activities reported by OMB increased from $4.6 billion to $8.8 billion (91 percent, or 62 percent after adjusting for inflation). Tax expenditures and funding provided in the Recovery Act are not included in this total to maintain consistency with our 2005 report on climate change funding.[13] In annual reports and testimony before Congress, OMB reported climate change funding for 1993 through 2010 using four categories:[14]

- *Technology*, which includes the research, development, and deployment of technologies and processes to reduce greenhouse gas emissions or increase energy efficiency. For example, according to OMB's reports on climate change expenditures, the activities counted in the technology category have the effect of stimulating the development and use of certain energy technologies, including renewable, low-carbon fossil, and nuclear technologies.
- *Science*, which includes research, modeling, and monitoring to better understand climate change; efforts to assess vulnerability to climate impacts; and programs to provide climate information to policymakers and the public. USGCRP also reports annually on funding for climate change science.
- *International assistance*, which helps developing countries to address climate change by, for example, providing funds for energy efficiency programs.
- *Wildlife adaptation funding,* which summarizes certain activities at the Department of the Interior designed to promote adaptation—adjustments to natural or human systems in response to actual or expected climate change. This category was described by OMB as an interim category in its June 2010 report because the Administration is developing criteria to systematically account for a broader suite of adaptation programs.

As shown in figure 1, technology funding increased as a share of total federal climate change funding from 2003 through 2010 (from 56 to 63 percent), continuing the trend observed in our 2005 report on climate change funding.[15]

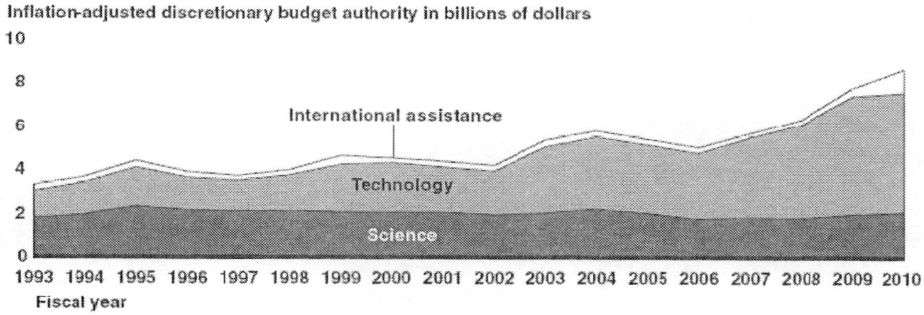

Source: GAO analysis of OMB data.

Notes: In its June 2010 report, OMB began reporting funding for wildlife adaptation as an interim category while criteria are developed to more systematically account for a broader suite of adaptation programs. Funding for wildlife adaptation totaled $65 million in 2010, less than 1 percent of the total, and is not included in this figure.

Funding in the American Recovery and Reinvestment Act of 2009 (Pub. L. No. 111-5 (2009)) and tax expenditures related to climate change are not included in this figure to maintain comparability with our 2005 report. Funding for these categories is presented separately in appendixes V and VI, respectively.

Figure 1. Reported Federal Climate Change Funding by Category, 1993-2010.

From 2003 through 2010, total technology funding increased from $2.56 billion to $5.5 billion (115 percent, or 83 percent after adjusting for inflation), according to OMB. Increases for Department of Energy (DOE) activities were the principal drivers of the growth in funding for technology activities, as they accounted for 80 percent of all technology activities in 2010. Increases within DOE programming were concentrated in the energy efficiency and renewable energy account.

In comparison, total funding for climate change science programs and activities remained relatively flat, and funding for international activities only recently increased substantially. Specifically, OMB reported that funding for climate change science activities and programs increased from about $1.77 billion in 2003 to $2.12 billion in 2010 (20 percent, or 2 percent after adjusting for inflation). Funding for international assistance activities decreased from $270 million in 2003 to $227 million in 2008 (15.9 percent or 27 percent after adjusting for inflation), but then increased significantly in both 2009 and 2010. Specifically, international assistance funding increased to $373 million in 2009, and increased further to $1.08 billion in 2010 (an increase of 376 percent over 2008 totals), according to OMB.

OMB's June 2010 report introduced several changes in report content and format. Specifically, OMB reported $65 million in funding for the interim category of wildlife adaptation activities. Appendixes III, IV, and V present climate change funding by category, agency, and account, as reported by OMB.

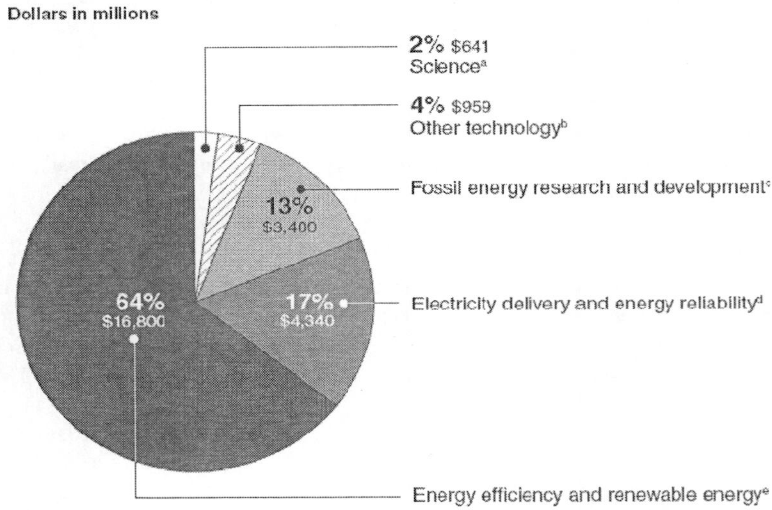

Source: GAO analysis of OMB data.

[a] *Science* includes funding for climate change research activities reported by OMB as USGCRP accounts and is presented in appendix V.

[b] *Other technology* includes funding for other Climate Change Technology Program (CCTP) accounts not presented in figure 2 as reported by OMB. These accounts are presented in appendix V.

[c] The DOE fossil energy research and development–efficiency and sequestration account includes efforts to ensure clean, affordable energy from traditional fuel resources, as reported by OMB and presented in appendix V.

[d] The DOE *electricity* delivery and energy reliability account includes activities to modernize the electric grid, and to enhance the security and reliability of energy infrastructure, among other things, as reported by OMB and presented in appendix V.

[e] The DOE *energy efficiency and renewable energy* account includes the research and development of renewable energy technology and efforts to improve energy efficiency, as reported by OMB and presented in appendix V.

Figure 2. Funding for Climate Change Programs and Activities in the Recovery Act as Reported by OMB.

Funding for Climate Change Programs in the Recovery Act

OMB reported $26.1 billion as funding for climate change programs and activities in the Recovery Act, dwarfing the $8.8 billion reported for similar activities in 2010.[16] About 98 percent of this funding was for climate change technology projects, while the remainder was for science activities.[17] We examined climate-related Recovery Act funding in several reports, including a review of DOE programs for innovative energy projects to help mitigate climate change.[18] Figure 2 shows the four largest categories of funding for climate change programs in the Recovery Act, as reported by OMB.

Recovery Act funding is presented separately in appendix V and is generally not included in summary tables cited in this report to maintain comparability with our 2005 report.

As Reported by OMB, Estimated Revenue Loss from Energy-Related Tax Expenditures Approached the Amount of Funding for Climate Change Programs and Activities in 2010

OMB also reports tax expenditures related to climate change, which are federal income tax provisions that grant preferential tax treatment to encourage emission reductions by, for example, providing tax incentives to promote the use of renewable energy.[19] In 2010, OMB listed 11 tax expenditures and energy grants with revenue losses totaling $7.23 billion, approaching the reported funding of $8.8 billion for climate change programs and activities. This sum includes $3.1 billion for energy grants in lieu of technology and investment tax credits available under the Recovery Act. Figure 3 illustrates selected tax expenditure categories in 2010, as reported by OMB.

The estimated revenue loss of $7.23 billion in 2010 from energy tax provisions that may reduce greenhouse gases is over 12 times the $580 million reported by OMB for 2003.[20] Reported revenue loss from tax expenditures increased steadily from 2006 through 2008, averaging about $1.5 billion. Tax expenditures reported by OMB then jumped substantially to $2.92 billion in 2009 and $7.23 billion in 2010, reflecting, in part, certain grants authorized in the Recovery Act. Climate-related tax expenditures are presented separately in appendix VI and are generally not included in summary tables cited in this report to maintain comparability with our 2005 report.

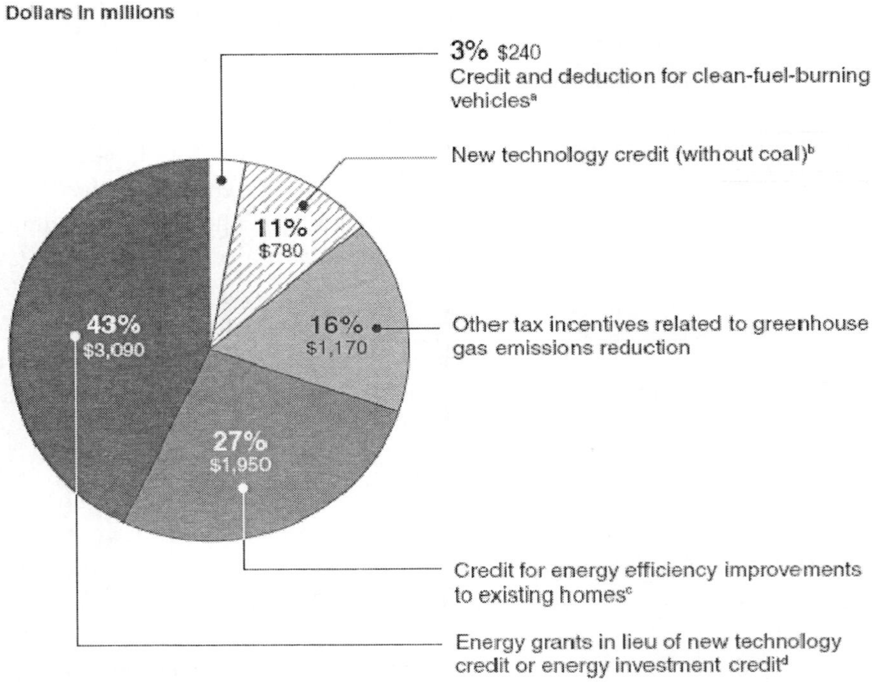

Source: GAO analysis of OMB data.

[a] *Credit and deduction for clean-fuel burning vehicles*: The tax code allows a number of tax credits for certain types of alternative motor vehicles (including fuel cell, advanced lean burn technology, hybrid, and alternative fuel motor vehicles) and other related technology.

[b] *New technology credit (without coal):* The tax code includes a tax credit for certain electricity produced from wind energy, biomass, geothermal energy, solar energy, small irrigation power, municipal solid waste, or qualified hydropower.

[c] *Credit for energy efficiency improvements to existing homes*: The tax code includes an investment tax credit for expenditures made on insulation, exterior windows, and doors that improve the energy efficiency of homes and meet certain standards.

[d] *Energy grants in lieu of new technology credit or energy investment credit* refers to Section 1603 of the Recovery Act, which authorizes the Department of the Treasury to make payments to persons who place in service specified energy property in 2009 and 2010 or whose construction commenced in 2009 and 2010. Firms can take an energy grant in lieu of the energy production or energy investment tax credit.

Figure 3. Estimated Revenue Loss from Energy Tax Provisions That May Reduce Greenhouse Gases as Reported by OMB, 2010.

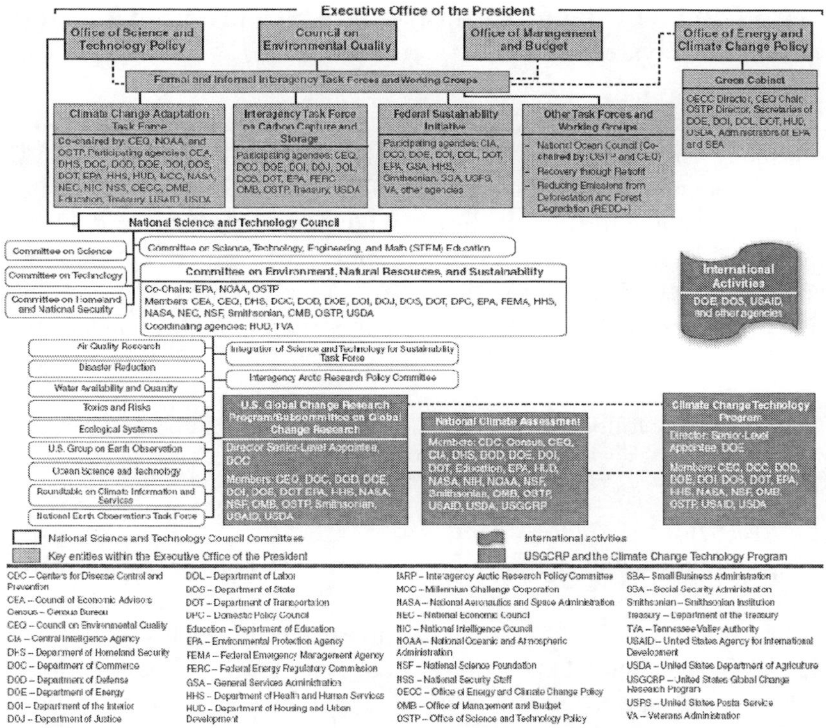

Source: GAO analysis of Climate Change Science Program Strategic Plan, Climate Change Technology Program Strategic Plan, NSTC organizational chart and CEQ initiatives.

Notes: The solid line from CEQ to the blue boxes indicates its role in administering these groups, and the dashed lines from OSTP, OECC, and OMB indicates their involvement. OSTP co-chairs, with NOAA and CEQ, the Interagency Climate Change Adaptation Task Force, and OSTP and CEQ jointly administer and co-chair the National Ocean Council.

According to technical comments from CEQ, OMB, and OSTP, figure 4 does not include all climate activities, such as individual agency regulations and initiatives. Some of these are described in more detail in appendix VII. CEQ, OMB, and OSTP also emphasized that informal processes and ad hoc meetings play key roles in setting climate priorities in addition to the formal processes described in figure 4. In technical comments, CEQ, OMB, and OSTP submitted table 3 in appendix VIII as supplemental information on how federal climate change activities are coordinated.

Figure 4. Selected Coordination Mechanisms for Federal Climate Change Activities.

Federal Climate Change Programs Are Organized in a Complex, Crosscutting System

As illustrated in figure 4, climate change is a complex, crosscutting issue, where many federal entities manage related programs and activities. These include organizations within the EOP (colored blue), interagency committees (colored white), and interagency programs (colored green) that coordinate the actions of individual agencies. A September 2010 report by the National Academy of Public Administration (NAPA), which was prepared for the National Oceanic and Atmospheric Administration (NOAA) and Congress, referred to this set of federal activities as the federal "climate change enterprise."[21]

The text that follows describes the roles of the entities involved in this enterprise.

Executive Office of the President

Four entities within EOP—represented by the blue boxes in figure 4—provide high-level policy direction for federal climate change programs and activities: OMB, OSTP, the Council on Environmental Quality (CEQ), and the Office of Energy and Climate Change Policy (OECC).[22] These entities lead several new interagency initiatives—including formal and informal task forces and working groups discussed later in this report— and participate in other committees, programs, and activities. For example, OMB is closely involved with budgetary decision making at USGCRP, the interagency body that coordinates federal climate change science research.

Interagency Committees and Programs

The National Science and Technology Council (NSTC)—represented, along with subordinate committees, as the white boxes in figure 4—coordinates science and technology policy across the federal government. NSTC is composed of four primary committees, including the Committee on Environment, Natural Resources, and Sustainability (CENRS), which coordinates federal research and development related to environment, natural resources, and sustainability. CENRS is composed of several subcommittees, one of which—the Subcommittee on Global Change Research—serves as the interagency governing body for USGCRP.

USGCRP and the Climate Change Technology Program (CCTP)—the green boxes in figure 4—coordinate federal climate change programs and activities at the agency level. Thirteen departments and agencies participate in

USGCRP, and 12 participate in CCTP.[23] CCTP and USGCRP coordinate through joint membership, according to technical comments from CEQ, OMB, and OSTP.

USGCRP, which began as a presidential initiative in 1989, was codified by the Global Change Research Act of 1990.[24] USGCRP coordinates and integrates federal research on changes in the global environment and their implications for society, and is led by an interagency governing body, the CENRS Subcommittee on Global Change Research (SGCR). The SGCR, facilitated by a national coordination office, provides overall strategic direction and is responsible for developing and implementing an integrated interagency program. The leadership structure for USGCRP falls within the SGCR, and includes a Chair and three vice-chairs: a Vice- Chair for Strategic Planning and Research, a Vice-Chair for Adaptation Science, and a Vice-Chair for Integrated Observations staffed by officials from different agencies. The committee leadership and agency representatives oversee and direct the program, including setting top-level goals and coordinating investments. CEQ, OMB, and OSTP noted in technical comments that OSTP provides overall leadership and direction by administering the NSTC and its committees.

CCTP was created in 2002 and subsequently codified in the Energy Policy Act of 2005. Its primary purpose is to assist in the interagency coordination of climate change technology research, development, demonstration, and deployment to reduce greenhouse gas intensity.[25] It provides strategic direction for climate change technology elements of the overall federal research and development portfolio, and facilitates the coordinated planning, programming, budgeting, and implementation of the technology development and deployment aspects of U.S. climate change policy. CCTP is managed by a DOE-led steering group composed of senior- level officials from each participating federal agency. This group provides a venue for agencies to raise and resolve issues regarding CCTP and its functions as a facilitating and coordinating body. The steering group assists in developing agency budget proposals and conveying information to agencies.

Congressional Committees

Not represented in figure 4 are congressional committees, which play a role in providing funding and oversight for climate change programs and activities. The size and scope of funding for federal agencies is determined through the federal budget process, which begins with budget formulation, then moves through the congressional budget process that includes consideration of appropriations legislation, budget execution and control, and finally audit and evaluation.[26] Appropriations bills are developed by the House and Senate Appropriations Committees and their subcommittees. Each subcommittee has jurisdiction over specific federal agencies or programs. For example, as a result of its interagency composition, activities of USGCRP participating agencies are usually funded by different appropriations bills that originate in nine separate committees.[27]

Questionnaire Responses Suggest That Agencies Do Not Consistently Interpret Methods for Defining and Reporting Climate Change Funding

Methods for defining and reporting climate change funding are not interpreted consistently across the federal government, according to questionnaire responses, available literature, and stakeholders. In responding to our questionnaire, federal officials identified three main methods for defining and reporting climate change funding, including formal guidance from OMB. Notwithstanding these efforts, questionnaire responses suggest that agencies do not consistently define and report climate change funding. While most respondents indicated that their own organization consistently interpreted and applied methods for defining and reporting climate change funding, far fewer said that other agencies across the federal government did so. Certain factors identified by respondents help to explain this limitation, including the difficulty in distinguishing between programs related and not related to climate change.

Respondents Identified Three Main Methods for Defining and Reporting Federal Climate Change Funding

In responding to our questionnaire, federal officials identified three main methods for defining and reporting climate change funding.[28] According to OMB staff, each of these methods has different levels of formality and oversight. First and foremost, agencies rely on guidance provided by OMB through Circular A-1 1, which describes how agencies are to report funding to OMB.[29] The circular directs agencies to report funding that meet certain criteria in three broad categories—research, technology, and international assistance. Budget data requested by OMB are reported by agencies through its Web-based MAX Information System (MAX) using the criteria specified in Circular A-11.[30] According to OMB staff, Circular A-11 is the primary methodology for defining and reporting long-standing "cross-cuts" of funding for climate change activities. Interagency groups, such as USGCRP have collaborated in the past with OMB to clarify the definitions in Circular A-11, according to technical comments from CEQ, OMB, and OSTP.[31]

Second, some respondents identified guidance from interagency programs as another method for defining climate change funding. For example, in its 2006 strategic plan, CCTP developed a list of classification criteria for how to report climate change technology funding.[32] According to a high-ranking CCTP official, this list was negotiated between its constituent agencies, and contains guidance about how to report funding. Similarly, USGCRP developed definitions to help agencies report research funding in categories related to its strategic research objectives. USGCRP has the sole responsibility for developing the definitions for more detailed science research categories, although the total spending reported by each agency should be consistent with the official figures released in the federal climate change expenditures report, according to technical comments from CEQ, OMB, and OSTP. These EOP entities also commented that funding analyses conducted by interagency programs are reviewed prior to release by OMB. Some agencies have also developed operational definitions of climate change programs within their purviews, according to several respondents. In technical comments, CEQ, OMB, and OSTP noted that such definitions help agencies ensure consistency over time in how they interpret criteria.

Third, entities within EOP or interagency programs may informally ask agencies to provide information on programs as part of periodic "data calls," according to several respondents. In technical comments, CEQ, OMB, and OSTP stated that data calls are most commonly used to collect information from agencies for analyses internal to EOP, but also for unique or new reporting needs. For example, USGCRP recently completed a data call in support of our September 2010 report on geoengineering.[33] The data call requested agencies to report federal funding for programs or activities that research deliberate, large-scale interventions in the earth's climate system to diminish climate change or its impacts, and it requested descriptions of specific geoengineering approaches that fit this definition. Similarly, OMB sent a recent data call asking agencies to report on programs or activities related to climate change adaptation.

Respondents identified two key strengths of current methods for defining and reporting climate change funding. The first was the use of stable processes, which enable an agency to compare its own climate change- related funding data over time, according to many respondents. As one stated, "we have a standardized process that accompanies the budget development process to collect climate change funding information.... Our agencies and offices are familiar with the process and are able to compile information relatively quickly." Another respondent noted that the strengths of current methods include a well-documented, systematic approach to budgeting that is based on solid requirements. The existence of a recurring process means that reporting can remain consistent across administrations, according to several respondents. According to OSTP, "there is a great deal of continuity in agencies' reporting of climate change programs and also continuity for most definitions of climate change... Most agencies' climate programs are ongoing, so it is relatively straightforward to report them year after year and also straightforward to review agencies' data every year."[34] In addition, some respondents noted that such continuity is useful in describing trends and can help inform Congress about priority federal activities.

Attention from high-level entities such as OMB is the second strength of current methods for defining and reporting climate change funding, according to several respondents. According to OSTP, climate change funding data may be reviewed three times each year for consistency and accuracy: in the OMB Circular A-1 1 process, the process for developing OMB's climate change expenditures report to Congress, and the USGCRP process for reporting climate science funding to Congress.[35] For the latter two reports, the data are presented in enough detail to allow for programby-program review and comparisons with agency budgets, according to OSTP.

Questionnaire Responses Suggest That Methods Are Not Consistently Interpreted and Applied across the Federal Government

Notwithstanding current methods, questionnaire responses suggest that methods for defining and reporting climate change funding are not consistently interpreted and applied across the federal government. While agencies generally exhibit continuity in how they assemble and report their own federal climate change funding data, methods for defining such funding are not consistent *across* the federal government, according to questionnaire responses. Respondents identified several factors that help to explain this inconsistency, including the number and complexity of climate-related programs managed by different agencies and the difficulty in distinguishing between programs related and not related to climate change. In technical comments, CEQ, OMB, and OSTP noted that consistency likely varies by method of reporting, with the USGCRP crosscut (as defined by Circular A-1 1) being the most consistent and "unique" data calls being less so.

Respondents Indicated That Agencies Interpret and Apply Existing Methods Differently

Respondents indicated that agencies do not consistently interpret existing methods for defining and reporting climate change funding *across* the federal government. Most respondents said that their agencies consistently applied methods for defining and reporting climate change funding.

However, far fewer respondents indicated that methods for defining and reporting climate change funding were applied consistently across the federal government.

Several respondents indicated and stakeholders said that federal agencies use their own interpretation of definitions to account for climate-related activities, resulting in an inconsistent accounting of these activities across the federal government. For example, according to one respondent, "agencies self-define their contributions, some use inclusive definitions ... while others use restrictive definitions ... and yet others do not report any investment. This makes it difficult to get a true picture of ... investment in climate change." The respondent also indicated that differing definitions do not "allow agency-to-agency comparisons since some agencies report apples while others report oranges." Another respondent stated, "to the best of my knowledge, nobody actually oversees a process for ensuring that the different agencies use similar methodologies for reporting." The issue of consistency also arises when federal departments collect climate change funding data from agencies within the department. For example, according to one respondent, "we have found inconsistencies within our own Department from year to year because of individual agency interpretations of the guidance. This has required significant discussion and even re-submissions to try to ensure consistency."

According to technical comments from CEQ, OMB, and OSTP, "despite the existence of common definitions, agencies must rely on their own interpretations. This leaves the perception, if not the reality, of inconsistencies in how agencies report climate change funding." According to OSTP, while reporting *within* agencies may be more or less consistent, there is not always a mechanism in place to ensure consistent reporting *across* agencies for each of the three methods for defining and reporting federal climate change funding. Such inconsistency is more applicable to data reported through interagency programs and unique data calls, according to CEQ, OMB, and OSTP. These EOP entities also said that respondents located within an agency may not have the proper vantage point to evaluate the internal processes in other agencies.

OMB staff told us that spending occurs at the subaccount level and that it must therefore rely in part on the agency to properly interpret the definitions. OMB staff stated that they are very responsive to instances where they learn agencies are over- or under-reporting official figures and were interested in specific examples of where there were inconsistencies in their various procedures. If there is a known inconsistency, OMB's procedure is to address it by discussing with the agency and clarifying Circular A-11 or relevant budget data request.

Respondents Identified Key Reasons Agencies May Interpret and Apply Existing Methods Differently

According to respondents, the following reasons may help explain why agencies could use different interpretations of the existing methods for defining and reporting climate change funding:

Size and complexity of the federal climate change enterprise. The overall scale of the federal climate change enterprise makes it difficult for officials to be aware of the whole range of programs and activities, according to respondents, available literature, and stakeholders. As noted by one respondent, instances have arisen where agency program staff are unaware or have not included funding data for certain programs to OMB, leading to confusion over how an agency's budget numbers were determined. According to OMB staff, some programs may not be reported as climate-related because agency officials may not have identified them in their budget reviews. For example, a high-level OMB staff member we interviewed told us that "reports may not be consistent if agencies don't report things they should. OMB tries to be consistent over time, but is dependent upon agency submissions."

Related to the current methods for defining and reporting funding is the practice of aggregating funding for multiple programs or activities without clear explanation. For example, one respondent said that it sometimes makes the accounting easier to lump non-climate-change activities in with climate change. Similarly, another federal official stated that the primary challenge associated with tracking climate change technology funding is that many programs can be included in a given appropriation line item. According to a 2008 CRS report, "because of the aggregation of funding information that is publicly available, understanding the specific uses of climate change funds can be challenging. The levels of funding for specific activities are often unreported or unclear. "[36]

Difficulty determining which programs are related to climate change. Determining "where to draw the line" between programs and activities related and not related to climate change is a limitation of current methods for defining and reporting funding, according to respondents, available literature, and stakeholders. Some respondents noted that it is difficult to make the distinction between what programs should and should not be counted as funding for climate change. Part of the difficulty lies in where to draw the line between "direct" programs and activities—for which climate change is a primary purpose—and "indirect" funding—which includes those programs that have a different primary purpose but support climate change goals.[37] The distinction is difficult to make, in part, because the primary purpose of direct funding may be to provide coherence to the disparate aims of the indirect programs, according to one stakeholder.

One respondent put it this way:

> "The biggest limitation... is separating what activities are explicitly 'climate change' vs. broader environmental activities which have impact on our knowledge of climate change. Our work relating to climate change frequently also relates to other environmental concerns—such as biodiversity patterns and loss, environmental dynamics and systems. Teasing out the 'climate' elements...is never easy or simple."

Indeed, such challenges of definition render the tallies of certain types of climate change programs somewhat arbitrary, according to a 2010 report by the Congressional Budget Office.[38] CRS raised similar concerns in a 2008 report, noting that "inconsistencies are likely across years due to changes in the scopes of what is considered for 'climate change.'"[39] Such inconsistencies make it harder for Congress and the public to fully understand how climate change funds are accounted for and how they are spent.

According to OSTP, if agencies reported all the investments in global change research (a much broader scope than climate), then the process would be more difficult, but possibly more comprehensive than the current approach.

OSTP added that the programs, and even the agencies, that are involved in climate issues are expanding rapidly as adaptation and mitigation activities are included, which makes definitional issues and reporting challenges even more difficult.

These views echo a long-standing definitional problem, identified as early as 1993 by the Congressional Office of Technology Assessment (OTA) in a report on climate change.[40] At that time, according to the report, climate-related projects were categorized as focused (directly related to global change) or contributing (justified on a basis other than global change but having the potential to contribute to the global change knowledge base). According to OTA, there were no standardized criteria for classifying contributing research, and each agency used its own system, making it harder to track overall climate change funding. Further highlighting the difficulty of defining climate change programs, a DOE official told us that "it can be difficult to distinguish between programs with direct climate change benefits and indirect benefits, especially for crosscutting items like basic research. If CCTP only reported those programs that were exclusively climate change-oriented, the number of programs and amount of funding reported would shrink dramatically. Many programs, such as wind energy and energy efficiency, have easily identified climate change co-benefits. However, there may be other programs with climate change co-benefits that are more difficult to identify, such as clean coal technology."[41]

Difficulty reconciling reporting categories with changing priorities. Some respondents noted that it is difficult to use standardized reporting categories when priorities change. For example, one respondent said the current methods are not flexible and do not allow for the inclusion of climate change activities that are very relevant but are happening outside of what is traditionally considered part of the climate change budget. Another respondent reported this was the case with adaptation activities, but also noted that agencies are in the process of developing definitions and reporting methods for the crosscutting, nontraditional category. One federal official we spoke with noted that, because of historical inertia, the current reporting process may not fit well with changing priorities. This official stated that "it is hard to change the reporting categories used in USGCRP's *Our Changing Planet* reports even though they no longer make any sense. It is risky to change such categories because people will ask questions about why the change was made."

According to several respondents, a recent proliferation of unique budget data inquiries have made it more difficult to consistently track and report funding data. According to one respondent, "too many different entities are requesting similar, but different information related to climate change." Another respondent said, "the way that we collect data and maybe even the data that we collect may inhibit our flexibilities in responding to the varying and increasing calls for information from outside entities. For example, our process may be designed to respond to OMB requests for information but may not directly respond to requests from other entities if the information requested is different." In technical comments, CEQ, OMB, and OSTP noted that respondents are likely referencing the difficulty in responding to unique data calls, not that the data calls undermine the effectiveness of long-standing crosscuts.

Agency support for USGCRP. According to OSTP, each agency pays its share of USGCRP's budget, which is determined by the amount of climate change research funding it reports. Hence, the more climate research funding an agency reports, the higher its assessment for supporting the USGCRP office and other joint responsibilities. According to technical comments from CEQ, OMB, and OSTP, this assessment was 0.35 percent in fiscal year 2010. Several stakeholders who were formerly involved in federal climate change funding decisions said that distributing the cost of funding USGCRP in this way provides an incentive for agencies to be cautious in how they report climate change activities. Some respondents echoed this concern. In certain situations, agencies may report conservatively to, for example, pay a lower assessment. In other instances, agencies may take the opposite approach. For example, according to one stakeholder, some agencies over-report their climate change activities because they see an opportunity to take credit for leadership.

OMB staff acknowledged this may be the perception, but disagreed that agencies could manipulate official climate spending figures. According to technical comments from CEQ, OMB, and OSTP, it is questionable whether official numbers can be manipulated in this way because OMB reviews annual variations to ensure that criteria are being consistently applied over time. More likely, respondents are referring to informal data calls conducted by EOP and interagency committees, according to the technical comments.

RESPONDENTS, AVAILABLE LITERATURE, AND STAKEHOLDERS IDENTIFIED TWO KEY FACTORS THAT COMPLICATE EFFORTS TO ALIGN CLIMATE CHANGE FUNDING WITH FEDERAL STRATEGIC PRIORITIES

Respondents, available literature, and stakeholders identified several existing mechanisms intended to align climate change funding with strategic priorities. These sources identified two key factors that complicate these efforts: (1) the lack of a shared understanding of federal strategic priorities among federal officials and (2) the fact that existing mechanisms that could help align agency funding with priorities are nonbinding, limiting their effectiveness where they conflict with agency responsibilities and priorities.

Existing Mechanisms Intended to Align Funding with Priorities

Respondents identified existing mechanisms intended to help align climate change funding with priorities. Several said that the budget development process aligns agencies' climate change funding with federal strategic priorities. For example, one said that funding is closely aligned with strategic priorities through the annual budget process, the research and development priority setting process, and interagency coordination meetings. Many also said that current interagency processes harness a broad array of stakeholders, and that their participation leads to stronger alignment of funding and priorities, both across the federal government and within their organizations.

OMB, OSTP, and CEQ use executive-level guidance memoranda to define climate change priorities within the overall federal budget. The Directors of OMB and OSTP described climate change priorities within the science and technology budget in a July 21, 2010, memorandum for the heads of executive departments and agencies titled *Science and Technology Priorities for the FY 2012 Budget*.[42] The memorandum instructs agencies to explain in their budget submissions how they will redirect available resources, as appropriate, from lower-priority areas to science and technology activities that address six challenges. One of the six challenges identified in the memorandum is "understanding, adapting to, and mitigating the impacts of global climate change." Specifically, agencies are requested to identify the activities in their budgets that support two priority areas—the National Climate Assessment (described in more detail in app. VII) and the monitoring, reporting, and verifying of greenhouse gas emissions. OMB and OSTP also issued supplemental guidance on climate change science collaboration on August 13, 2010.[43]

The Directors of OMB and OSTP issued a similar memorandum for the fiscal year 2011 budget on August 4, 2009.[44] According to OSTP, overarching climate change priorities are also discussed by OECC policy committees and through discussions convened by the Assistant to the President for Energy and Climate Change.

Other processes for setting priorities are collaborative exercises between entities within the EOP, interagency coordinating programs, and agencies, according to most of the respondents who answered the related survey question. For example, as stated by one respondent,

> "the USGCRP and to a lesser extent the CCTP have responsibility for priority setting on climate change science. USGCRP is overseen by OSTP and the NSTC. The Council on Environmental Quality is responsible for coordinating federal environmental efforts across agencies and EOP offices, including development of policy and initiatives. Congress has responsibility for developing and enacting legislation including budget legislation that may include specific national, regional or state initiatives that may be conducted through federal agencies. National Academies / National Research Council is tasked with producing independent reports on successes and challenges to keep the process moving forward and identify obstructions to achieving strategic goals."

Many other respondents also noted the collaborative nature of climate change priority setting within the federal government. According to OSTP, "priorities listed in memos and documents that describe Administration policy ... are set by multiple iterations of discussions among EOP offices and the USGCRP principals. These reflect a broad process of Science and Technology priority setting, along with budget discussions, within the National Science and Technology Council..., in particular the Committee for Environment, Natural Resources, and Sustainability, of which the Subcommittee on Global Change Research is a part."

Several respondents noted that USGCRP recently launched a new strategic planning process, in part to improve how climate change priorities are set. For example, one respondent stated that "the USGCRP is undergoing a strategic realignment to become an integrated, end-to-end program. A strategic planning process has been initiated recently." Additional respondents noted that the program is at a turning point, having just launched a new planning process with greater emphasis on providing information and guidance for decision makers. CEQ, OMB, and OSTP technical comments noted that USGCRP also recently launched a new national climate assessment designed to engage stakeholders in a process that builds on science, data, and information to help decision-making for analysis and mitigation.

Questionnaire Responses Indicated That Federal Officials Lack a Shared Understanding of Priorities

Despite existing coordination mechanisms, federal officials across the climate change enterprise lack a shared understanding of priorities, according to questionnaire responses, available literature, and stakeholders. This is partly due to the multiple, often inconsistent messages articulated in different forums and in different policy documents. Our review of these sources found that there is not currently a consolidated set of strategic priorities that integrates climate change programs and activities across the federal government.

When asked what the priorities across the federal government were, the respondents who answered the question provided a wide range of answers. While the broad category with the most responses was "adaptation," a number of other categories—such as priorities listed in the USGCRP and CCTP strategic plans—were almost as frequently cited. Other responses included a new focus on providing climate change information to decision makers. For example, according to one respondent, "a recent emphasis seems to be connecting the use of [climate change] science to meet end-user needs....As a result, 'science translation' has become a new priority as well as advancing impact assessment and adaptation models and tools." Only a few respondents suggested that there was an absence of strategic priorities, with one stating, "I'd argue that there are not [any], or perhaps few, strategic priorities across government. Agencies have their own internal missions and priorities; I've seen little willingness or ability to shift those priorities and programming decisions to meet overall strategic interests, even following transition to a new Administration."

These results are supported by a 2008 CRS analysis, which states "there has not been an overarching policy goal for climate change that guides the programs funded or the priorities among programs. U.S. federal policy on climate change has been a coalescence of separate goals—evolved distinctly for science, technology, energy production, foreign assistance, and trade—not a single, integrated strategy. The current federal effort largely has been built 'bottom up' from a variety of existing programs, Presidential initiatives, and Congressionally-directed activities. Choices tend to be based on departmental missions and the degree of support for the input activities."[45]

In the absence of clear, overarching priorities, federal officials are left with many different sources that present climate change priorities in a more fragmented way.

Through our review of questionnaire responses, available literature, and interviews with stakeholders, we found that federal strategic climate change priorities are presented in six general sources: (1) strategic plans for interagency programs and agencies, (2) executive-level guidance memoranda, (3) the development of new interagency initiatives, (4) regulations and guidance memoranda, (5) international commitments, and (6) testimony of federal executives before Congress. Each of these sources is discussed in more detail in appendix VII.

The multiple sources for communicating priorities across the climate change enterprise may result in conflicting messages and confusion. Our review found multiple plans with different scopes, goals, and time frames. While OSTP and OMB most recently directed agencies to focus on activities in their budgets that support the national assessment and monitoring of greenhouse gas emissions, the 2003 USGCRP strategic plan and its 2008 revision generally focus federal research on different priorities, mostly related to reducing scientific uncertainty about climate changes.[46] CCTP's 2006 strategic plan similarly focuses federal resources on different priorities, including capturing and sequestering carbon dioxide. While these strategic plans were created for different purposes and times, and are not mutually exclusive, they—and the programs charged with implementing the plans—represent the operational link between EOP and agencies. According to a high-ranking OMB staff member we interviewed, "to the extent that OMB receives direction from the administration, it will implement the president's priorities. If that doesn't happen, OMB will take the existing strategic plans as direction."

Several respondents said there is no single lead to direct and coordinate federal climate change efforts, which nearly matches an observation from our 1990 report that "the President has not designated any individual or agency to assume overall leadership or management responsibility for global climate change."[47] For example, according to one respondent, there is "no single lead person to direct and coordinate all the efforts." Another respondent said "without clear direction and leadership by the White House, it is very difficult for the agencies to set strategic priorities." According to one stakeholder, "there are many willing partners in the federal government, but there is a small bottleneck at the top. Agencies want to take action but don't know who is in charge of administration policies."

The lack of a shared understanding of strategic priorities combined with the crosscutting nature of the federal climate change enterprise is likely to complicate efforts to align funding with priorities. Our past work on interagency collaboration has shown that collaborating agencies must have a clear and compelling rationale to work together to overcome significant differences in agency missions, cultures, and established ways of doing business.[48] This requires agency staff working across agency lines to define and articulate the common outcome or purpose they are seeking to achieve that is consistent with their respective agency goals and mission. Without a clear common purpose, it is more difficult for agencies to define roles and responsibilities within the complex climate change enterprise previously illustrated in figure 4. Agreeing on roles and responsibilities—clarifying who will do what—and committed leadership by those involved in the collaborative effort are necessary to overcome barriers to working across agencies.

Existing Mechanisms Intended to Align Funding with Priorities Are Nonbinding, Limiting Their Effectiveness Where They Conflict with Agency Responsibilities and Priorities

Existing mechanisms intended to align funding with strategic priorities are nonbinding, according to respondents, available literature, and stakeholders. For example, one respondent said that existing "coordination mechanisms are non-binding. It is difficult to reconcile the different missions of the agencies without strong commitment from each agency." Another respondent stated, "Undertaking interagency projects in climate change is extremely challenging. The system is not designed to encourage interagency activity... Each agency has its own processes, deadlines, and requirements." Other respondents noted that the interagency policy process does not control agency budgets, and that agencies with their own budget authority may pay little attention to federal strategic priorities. In other words, federal strategic priorities set through an interagency process may not be reflected in budget decisions for individual agencies.

Several federal officials we interviewed said that interagency coordinating programs like USGCRP generally do not have direct control over agency budgets. A high-ranking USGCRP official told us that "USGCRP does not control any resources. It doesn't do anything but put out reports. Program office staff help different agencies collaborate, but it doesn't amount to prioritization."[49] Similarly, a high-ranking CCTP official told us that every constituent agency of CCTP has control over its own budget; CCTP has no authority to direct funding for climate change technology programs. According to a 2009 National Research Council (NRC) report, the absence of centralized budget authority limits the ability of the USGCRP to influence the priorities of participating agencies or implement new research directions that fall outside or across agency missions.[50] The current approach for integration is likely to continue as long as strategic decision making is decoupled from agency budgetary processes or while USGCRP has no budgetary authority, according to observations in a 2008 USGCRP expert roundtable discussion.[51]

Several respondents observed that the number of staff available to coordinate governmentwide climate change programs is currently limited, which may also make it more difficult to align funding with priorities. For example, according to one respondent, USGCRP has a small number of staff, which limits the effectiveness of coordination efforts. Another respondent stated that the current model of a very small USGCRP staff office is not effective. A 2008 CRS report noted that "one potentially important element in the success of programs is the expertise of federal officials in these programs, and whether federal policies enhance or hinder the recruitment, development and effective use of personnel. Barriers to job mobility of federal personnel across programs and departments also likely discourage development, interaction and collaboration across agencies and disciplines." When provided with this information, a USGCRP official stated that "the USGCRP staff is very well qualified to do their job. This issue is not about qualification but about the lack of empowerment and respect from Federal Agencies."

Nonbinding mechanisms for aligning funding with priorities have limited effectiveness where they conflict with agency responsibilities and priorities, according to certain stakeholders and respondents who commented on the subject. According to them, funding is not aligned well with new priorities across the federal government because there is a reluctance to make room in existing budgets for new work or because most funding is pre-committed to existing programs, limiting the capacity to make major new funding commitments. For example, according to one respondent, "agency mission requirements can take precedence, and this can limit the effectiveness of responding to strategic priorities. This becomes increasingly the case when going from the highest leadership levels down into lower levels of the organization." Another respondent stated that her agency, "like all the agencies, has a very difficult time changing agency priorities and strongly resists stopping any current programs." As noted by a different respondent, agencies generally have to stop doing existing good things to do new good things, even when existing programs have not completed their objectives.

RESPONDENTS SUGGESTED OPTIONS FOR BETTER ALIGNING FEDERAL CLIMATE CHANGE FUNDING WITH STRATEGIC PRIORITIES

Respondents, stakeholders, and available literature identified several ways to better align federal climate change funding with strategic priorities. They suggested (1) options to improve the tracking and reporting of climate change funding, (2) options to enhance how strategic climate change priorities are set, (3) the establishment of formal coordination mechanisms, and (4) continuing efforts to link related climate change activities across the federal government. Federal entities reported that they are already taking steps to implement several of these options.

Options to Improve the Tracking and Reporting of Climate Change Funding

Respondents identified two key options to improve how climate change funding is tracked and reported, a key element in improving how funding is aligned with priorities: an integrated budget review process that provides a more centralized and predictable approach for collecting and reviewing climate change funding, and enhanced guidance on how to define and report funding. In technical comments, CEQ, OMB, and OSTP stated that they are already taking steps to implement these options.

An Integrated Budget Review Process

Many respondents called for a more coordinated, centralized approach to reporting and reviewing climate change funding—referred to by some as an integrated budget review process. Such a process could address the limitations of current methods for defining and reporting climate change funding, such as the difficulty in distinguishing between programs related and not related to climate change. Several respondents stated that requests for climate change funding data should come from one source to increase the efficiency and consistency of reporting. According to one respondent, "if possible, it would be great if one organization (OMB) were the sole entity that requested information from Departments. That would help to ensure consistency and the use of common definitions and terms." Another respondent added that "a more formal request—with better guidance on what should be reported—coming from OMB ... might impose more discipline and accuracy on the process." Other respondents noted that a more interactive relationship with OMB would help provide consistency in climate change funding reports. As we said earlier, OMB's MAX A-11 data entry system provides a centralized collection point for agencies to enter climate change budget data specified in OMB Circular A-11.

An integrated budget review process could also be a way to better align agency funding and interagency priorities.[52] According to OSTP, more interaction between OMB, OSTP, and agency budget and programmatic leads would help to develop more consistent reporting as well as provide a better framework for developing initiatives and building crosscut strategic elements. As a means of achieving this goal, OMB and OSTP held a budget hearing with USGCRP and agency officials in September 2010, according to OSTP. According to OMB and OSTP guidance on climate change science collaboration dated August 13, 2010, participating agencies were to

> "Prepare for a joint budget hearing in late September in which agencies, through their USGCRP Principals, articulate their climate change science request to OMB, OSTP, and other Executive Offices, including CEQ, and other USGCRP agencies. Agencies should coordinate prior to the meeting in order to present FY 2012 climate change science initiatives in appropriate detail, recognizing the relative size of agency investments to the National Climate Assessment and overall climate change science budget. Agencies should also address how their FY 2012 activities will build upon efforts by partner agencies while minimizing overlap and duplication."

OSTP cautioned, however, that there are inherent limitations in the system because priorities that span the federal government, by definition, involve multiple agency budgets. Communication is critical, and it is useful to have OMB, OSTP, and USGCRP talking to each other about priorities in the process of strategic planning and budget decisions, according to OSTP.

The joint budget hearing proposed last August by OSTP and OMB appears similar to past practices identified as useful by stakeholders. Several stakeholders we interviewed described how the federal climate change enterprise—successfully, in their view—integrated funding with priorities in the early 1990s. For example, according to one stakeholder,

> "USGCRP held a series of interagency meetings to develop the crosscut climate change science budget before sending it to OMB for review. At the meetings, each of the 13 agencies that were part of USGCRP had to prove their climate related activities were linked to the goals of the program. The integrated budget review process challenged people to match the priorities of the program in order to obtain funding. The review process was challenging, and probed the 'readiness' and relevance of individual programs before they received funding. The integrated budget reflected the best set of priorities at the time. An integrated budget review hasn't happened in 12 years."

Other stakeholders and sources said an integrated budget review would be useful because, in the past, such a system engaged agencies in the interagency process and fostered agreement on priorities.

OMB staff cautioned that there are trade-offs with a more formal process such as an integrated budget review, noting that "formality only works with a limited number of topics because it requires a great deal of effort ... . Over time, the process didn't work out very well because large meetings with high level officials are hard to schedule and take a lot of time." In addition, past integrated budget reviews focused solely on climate change science programs and activities because USGCRP was the only interagency program in existence. It is not clear how—or if—formal budget review processes could be applied more comprehensively across the climate change enterprise.

Additional Guidance from the Executive Office of the President on Defining and Reporting Federal Climate Change Funding

Many respondents reported wanting additional guidance from OMB, OSTP, or USGCRP about how to define and report climate change funding. As stated by one respondent, "Interactive discussions between responsible parties from OSTP, OMB, and [USGCRP] with the purpose of defining the true scope of funding for climate change would be very useful. Since areas of climate research change annually, this would be an ongoing process." Another respondent asked for a review and revision by OMB of all climate change reporting requirements to better define the categories and what information is needed. One approach could be to reach agreement with congressional appropriators on a set of definitions and criteria for climate change programs that could be used for several years, according to one respondent. In any case, several respondents noted that guidance needs to clearly articulate what is and is not considered to be climate change funding.

Options to Enhance How Strategic Climate Change Priorities Are Set

As we have previously reported, enhanced priority setting processes can encourage agency collaboration by defining and articulating a common outcome, reinforcing accountability for collaborative efforts through plans and reports, and establishing mutually reinforcing or joint strategies to achieve the outcome.[53] Respondents, available literature, and stakeholders identified several options for enhancing the process of setting federal climate change priorities, including (1) using a governmentwide strategic planning process, and (2) developing a clear leadership and coordination structure.

A Governmentwide Strategic Planning Process

According to questionnaire respondents, available literature, and stakeholders, a governmentwide strategic planning process could enhance how strategic priorities are set by articulating what individual agencies are expected to do within the overall federal response to climate change.[54] Some respondents noted the need to expand planning beyond individual elements—such as science or technology—and to develop a truly crosscutting climate change strategic plan. For example, according to one respondent, "there is a need now to go to the next level and coordinate across disciplinary areas in order to address the urgent challenges related to climate variability and change." Other respondents said that developing a more detailed and actionable list of priorities through existing planning processes would be helpful.

A number of federal climate-related strategic planning processes are under way or have been recommended. For example, USGCRP is undertaking a strategic planning process to ensure the alignment of Administration priorities, NRC guidance, and agency programs and budgets, according to CEQ, OMB, and OSTP technical comments. In addition, a 2010 NAPA report recommended that the President empower a senior interagency group—led at the White House and convened at the Deputy Secretary or Secretary level—to provide the President annually with a strategic plan for management of federal climate research and service delivery.[55]

As OSTP acknowledged to us, "The major challenge is the need to connect climate science programs with broader inter- and intra-agency climate efforts." OSTP stated that while significant progress is being made in linking the climate science-related efforts, individual agencies still want to advance initiatives that promote or serve their agency missions. This, according to OSTP, yields a broader challenge of tying climate-related efforts (science, mitigation, and adaptation) together into a coherent governmentwide strategy.

A Clear Leadership and Coordination Structure

Some respondents noted that it would be helpful to clearly identify a lead person or organization responsible for developing and integrating strategic climate change priorities across the government. According to several respondents, developing federal strategic climate change priorities would be easier if there was a clear leadership and coordination structure responsible for setting governmentwide climate change priorities. They cited a particular need to integrate policy, science, mitigation, adaptation, technology and the other aspects of the federal climate change enterprise into a single structure for setting priorities. One official stated that this would require "a real commitment to a full-time leader with staff and adequate resources to work closely with all of the Federal Agencies, USGCRP, OMB and the Hill in developing a strongly coordinated global change crosscut that has consistent goals and objectives with well-defined and linked performance measures, as well as the requisite authorities to make (or at least propose) changes to programs and budgets in order to better meet the measures and achieve the overall strategic goals and objectives."

Several respondents specifically cited the need to integrate USGCRP's focus on science with CCTP's focus on technology. As stated by one respondent, "until now, there has been a general disconnect between USGCRP and CCTP—we should connect the science and technology programs much more closely." According to OSTP, the USGCRP's strategic planning process will incorporate better mechanisms for linking with the CCTP. It maintains that in this way, climate technology plans, policies, and projections developed by CCTP can be integrated with those of the climate science and adaptation groups.

Several respondents did note that developing an overarching framework for setting priorities would be challenging. One respondent cautioned that the mix of technology, economics, and policy expertise cannot be found completely in one agency or EOP entity. As stated by another respondent, “This is bigger than any single agency. Therefore, no single agency can be put in charge.” Several respondents expressed a similar sentiment.

The Establishment of Formal Coordination Mechanisms

Respondents noted that the establishment of formal coordination mechanisms—such as executive-level memoranda that specify how agencies are to work together—could improve how federal climate change funding aligns with strategic priorities. Many respondents called for improved coordination to better align funding with priorities. Several identified related options, including formally identifying a single contact point for coordinating climate change funding within each agency and defining the role of this contact within interagency structures. For example, one respondent stated, “In each agency there needs to be a formal internal process that links budget (CFO) to strategy (Policy), reporting to the agency Head or Deputy.... Each agency team, in turn, needs to be linked to an interagency coordinating body that champions strategic planning, strengthens data and tools for decision-making, shares information with agencies, educates new managers and appointees, and presents integrated reports on status and progress toward goals.” In addition, several stakeholders we interviewed described how, in the early days of USGCRP, formal “terms of reference” defined who did what and the relationships between federal agencies and the EOP. The terms of reference included specific milestones of what would be produced by when, goals, roles, and responsibilities for each agency that was part of USGCRP.

A 2010 NAPA report noted that the development of a senior-level federal interagency coordination mechanism with a broad mandate would be enormously valuable, not only to coordinate federal climate research and service delivery, but also to potentially coordinate other climate initiatives.[56] In its report, NAPA recognized that the same senior interagency group needed to coordinate federal climate research and service delivery ideally could be configured to support the overarching executive branch climate policy agenda, which, according to the report, would likely bring beneficial synergy and focus to decision making.

Another approach for aligning funding with priorities that enjoyed success in the past, according to the results of a 2008 USGCRP expert roundtable discussion, was OMB’s past practice of favoring interagency priorities over individual agency priorities when collectively reviewing climate science budget decisions.[57] The roundtable discussion emphasized that there needs to be a "carrot" (a flexible pool of funding to encourage agency participation) and "stick" (a penalty for not addressing interagency priorities) to enhance agency implementation of interagency program priorities. According to our interviews with stakeholders, past USGCRP funding decisions vetted by an integrated budget review process, were enforced by what was known as "fencing." Once agencies signed off on their contribution to the USGCRP during budget negotiations with OMB, funding became "fenced off." Therefore, once an agency committed funds it was not allowed to "change its mind," as OMB would not allow it to reprogram the funds for other purposes elsewhere in its budget.[58] In other words, once the priorities were set through the budget review process, they were locked in through "fencing."

A different idea to improve the coordination of climate change programs suggested by several reports we reviewed was to provide USGCRP with a central budget to allocate to its member agencies, or the authority to redirect agency funding to interagency priorities. A 2009 NRC report stated that an increased discretionary budget for the USGCRP director, sufficient to carry out interagency efforts such as workshops would provide flexibility and seed money for objectives that are of higher priority to the program than to any participating agency.[59] However, certain respondents warned that giving budget authority to USGCRP was not practical, noting that the discretion that agencies have is appropriate because many climate change activities also serve other functions. Furthermore, according to the testimony of a former director of OSTP, "it is a reality that central budgeting for an interagency effort like [USGCRP] or CCTP is incompatible with the Federal budget structures and processes."[60]

EOP offices have begun taking steps to improve coordination. For example, a July 2010 memorandum from OMB and OSTP to department and agency heads detailing certain climate-related priorities states that "in requesting funds for large-scale [science and technology] projects involving significant interagency or international collaboration, agencies should identify: the lead organization for the collaboration; the unique capabilities brought to the collaboration by each partnering organization; and the specific roles and responsibilities for each organization. Agencies should coordinate with partner Federal agencies to formulate budget requests for interagency collaborations"[61]

Continuing Efforts to Link Related Climate Change Activities

Efforts to link related climate change activities across the federal government could better align climate change funding with priorities, according to respondents, available literature, and stakeholders. For example, developing a small number of tools that can serve all agencies would improve the government's overall effectiveness, according to one respondent. An area of possible linkage frequently mentioned by respondents includes efforts to coordinate climate services across the federal government. According to NOAA, a climate service could provide a single source for climate data, information, and decision support services to help individuals, businesses, communities, and governments make smart choices in anticipation of future climate changes. Several stakeholders noted that this service could better focus climate efforts to achieve federal strategic priorities. The 2010 NAPA report discusses the factors needed for a NOAA Climate Service to succeed—such as the designation of a lead federal agency to be the day-to-day integrator of the overall federal effort regarding climate science and services—and makes recommendations on how to achieve those factors.[62] Also in 2010, the NSTC's CENRS established the Roundtable on Climate Information and Services in recognition of the fact—stated in testimony and in National Research Council reports—that no single agency possesses all the information and capabilities of climate services delivery, according to technical comments from CEQ, OMB, and OSTP.[63]

There has been some recent progress on linking related federal climate change programs, according to OSTP. Specifically, OSTP stated that the science portion of the CEQ, NOAA, and OSTP-led Climate Change Adaptation Task Force is being integrated within USGCRP. OSTP also stated that it is working to create an interagency body that will bring together agencies that provide climate services to allow for better links between climate services and other federal climate-related activities.

CONCLUSION

Climate change has proven to be among the most challenging environmental issues currently being addressed by the federal government. Our work on similarly complex and interdisciplinary issues underscores the importance of close interagency coordination and collaboration to ensure that funds are properly accounted for and are spent so that they have their intended effect. This work also emphasizes that having agencies agree on roles and responsibilities—clarifying who will do what—is necessary to overcome barriers to working across agencies. The responses of federal officials to our Web-based questionnaire, together with other information gathered for this review, suggest that various issues may be limiting the effectiveness of the climate change enterprise. The questionnaire responses provide the unfiltered views of the federal officials who work directly on these issues and represent a range of vantage points on federal efforts. We worked with CCTP, CEQ, OMB, OSTP, and USGCRP to identify the officials most knowledgeable about governmentwide efforts to align funding with priorities, including respondents from entities within EOP, interagency coordinating programs, and individual agencies. Among the key issues identified by questionnaire responses, available literature, and interviews with stakeholders are that (1) methods for defining and reporting climate change funding are not interpreted consistently across the federal government, and (2) federal officials do not have a shared understanding of strategic priorities. The Administration has recognized the need for improvement, and has taken steps to both begin to augment the reporting of climate change funding and to better align funding with priorities. That said, officials responding to our questionnaire provided additional options that may further improve how agencies, interagency bodies, and entities within EOP define and report federal climate change funding, set strategic priorities, and align funding with priorities. Without further improvement in these areas, it will be difficult for Congress and the public to fully understand how climate change funds are accounted for and how they are spent.

Recommendations for Executive Action

To improve the coordination and effectiveness of federal climate change programs and activities, we recommend that the appropriate entities within the Executive Office of the President, including the Council on Environmental Quality, Office of Energy and Climate Change Policy, Office and Management and Budget, and Office of Science and Technology Policy, in consultation with Congress, work together with relevant federal agencies and interagency coordinating bodies to take the following two actions:

- Clearly establish federal strategic climate change priorities, including the roles and responsibilities of the key federal entities, taking into

consideration the full range of activities within the federal climate change enterprise.
- Assess the effectiveness of current practices for defining and reporting federal climate change funding and aligning funding with priorities, and make improvements to such practices as needed for Congress and the public to fully understand how climate change funds are spent.

Agency Comments and Our Evaluation

We requested comments on a draft of this report from the Chair of CEQ, the Director of OMB, and the Director of OSTP. They did not provide official written comments to include in our report. Instead, they provided technical comments, which we incorporated as appropriate.

Sincerely yours,

David C. Trimble
Acting Director
Natural Resources and Environment

Appendix I. Scope and Methodology

This report examines (1) federal funding for climate change activities and how these activities are organized; (2) the extent to which methods for defining and reporting climate change funding are interpreted consistently across the federal government; (3) federal strategic climate change priorities, and the extent to which funding is aligned with these priorities; and (4) options, if any, available to better align federal climate change funding with strategic priorities.

To collect information for these objectives, we analyzed climate change funding data and tax expenditures presented in Office of Management and Budget (OMB) reports, reviewed other relevant reports, and interviewed stakeholders knowledgeable about federal climate change funding programs and activities. To identify relevant reports, we reviewed our prior climate change work and conducted a literature search and review with the assistance of a technical librarian. To identify stakeholders, we reviewed our prior climate change work and relevant reports to identify individuals with specific knowledge of federal climate change funding. The stakeholders we interviewed were formerly involved with climate change funding decisions within the federal government.

Using information we gathered through our literature review and initial interviews with stakeholders, we developed a Web-based questionnaire to gather information and opinions of key federal officials involved in defining and reporting climate change funding, developing strategic climate change priorities, or aligning funding with strategic priorities. We designed the questionnaire to collect information through open-ended questions, organized in different sections to allow respondents to answer the questions most in line with their knowledge and experience, and to skip sections with which they were less familiar. The questionnaire was divided into five sections: (1) background, where we asked for a range of descriptive information about the respondents; (2) defining and reporting federal climate change funding; (3) setting federal strategic climate change priorities; (4) aligning federal climate change funding with strategic priorities; and (5) conclusion, where we asked for additional comments.

After we drafted the questionnaire, we asked for comments from select federal officials involved with climate change issues within the Executive Office of the President, interagency coordinating programs, and individual agencies. We conducted pretests with these officials to check that (1) the questions were clear and unambiguous, (2) terminology was used correctly, (3) the questionnaire did not place an undue burden on agency officials, (4) the information could feasibly be obtained, and (5) the questionnaire was comprehensive and unbiased. We made changes to the content and format of the questionnaire based on the feedback we received. A copy of the Web-based questionnaire is included in appendix IX.

To identify questionnaire respondents, we requested and received lists of knowledgeable officials from the Council on Environmental Quality (CEQ), Climate Change Technology Program (CCTP), OMB, Office of Science and Technology Policy (OSTP), and United States Global Change Research Program (USGCRP). On July 26, 2010, we sent an e-mail notifying 147 officials of their opportunity to participate in the questionnaire and described the topics that the questionnaire would cover. We asked these recipients to notify us if they were not the appropriate person to fill out the questionnaire, and to tell us who might be appropriate, if possible. Based on these responses, we removed 11 officials from our list of potential respondents and added 3. Consequently, we invited a nonprobability sample of 139 officials to complete our Web-based questionnaire. We later removed an additional 33 officials from the respondent list because we learned, through follow up phone calls or e- mails, that certain individuals were either not knowledgeable or not available (e.g., no longer federal employees). The final list of potential respondents included 106 federal officials.

We administered a Web-based questionnaire that was accessible through a secure server. On August 6, 2010, we sent an e-mail announcement to notify the respondents that the questionnaire was available online, and they were given unique passwords and usernames. We sent a follow-up email message on August 27, 2010, to those who had not yet responded. Then we contacted all remaining nonrespondents by telephone, starting on September 7, 2010. The questionnaire was available online until September 24, 2010.

Of the 106 officials who were asked to participate, 73 responded to the questionnaire, for a response rate of about 69 percent. Not all officials responded to all questions. The officials who responded hold a variety of positions within the federal government and represent a diverse array of disciplines. For example, when respondents were asked to best describe their positions within the federal government, 7 responded "budget- oriented," 9 responded "policy-oriented," 15 responded "both budget- and policy-oriented," 19 responded "program manager," 7 responded "scientist," and the rest did not answer or responded "other." Respondents were from entities within the Executive Office of the President, interagency coordinating programs, and individual agencies. Given our methodology, we may not have identified every person who has knowledge of or experience with the topics we covered; however, we do believe we were able to reach many of the relevant federal officials because we worked with CCTP, CEQ, OMB, OSTP, and USGCRP to identify potential participants.

We asked several OSTP officials to respond individually to the questionnaire, but OSTP elected to provide a single formal response. OSTP officials stated that their collective response was the subject of lively internal debate and that they believed that questionnaire responses from individual federal officials that had not been vetted by agency management deserved less weight. Throughout this report we attribute information from this formal collective response directly to OSTP using phrases such as "according to OSTP" or "OSTP stated."

The questionnaire presented in appendix IX asked a series of open-ended questions. As already noted, not all officials responded to all questions. We analyzed questionnaire responses by developing and testing a coding structure and coding the interviews using qualitative analysis software. We used this software to help us group similar responses together into overall themes to assist in the writing process. We did not use the software for statistical analysis because the sample of respondents was not a representative sample.

We used specific responses as illustrative examples throughout the report, and employed a simple scale to describe the extent to which respondents made statements related to a theme or other topic discussed in the report. Descriptors are in relation to the total number of respondents who commented on a particular theme or topic, which varied by the theme or topic. We use "several," "certain," and "some" interchangeably to mean three or more, but fewer than one-third. We use "many" to mean more than one-third, and "most" to mean more than half of the relevant respondents. If, in the report, we make a general statement like "respondents said...", then the text following the statement will provide additional details. To make the report more readable and less confusing, we do not include the numbers of respondents who commented on particular themes.

We received and incorporated comments from federal entities. On December 17, 2010, CEQ, OMB, and OSTP submitted consolidated technical comments. We accounted for these comments in our draft report. We also incorporated technical comments submitted from CEQ, OMB, and OSTP on April 27, 2011.

The term "funding" in this report reflects budget authority, or the authority provided by federal law to enter into financial obligations that will result in outlays of federal government funds, as reported by OMB. Unless otherwise stated, we report funding in nominal terms (not adjusted for inflation), and all years refer to fiscal years. Totals and percentages may not add due to rounding. When we adjusted for inflation, we used a fiscal year chain weighted gross domestic product price index composed of averages of quarterly indexes from the U.S. Department of Commerce, Bureau of Economic Analysis, *National Income and Product Accounts*, table 1.1.4. Unless otherwise specified, figures represent actual funding (not estimates), with the exception of 1993, 1994, and 2010, where we present estimated funding because actual data are not available.

For the purposes of this report, the term "agency" includes cabinet-level departments and other agencies, and we use the term "account" to describe the budget accounts, line items, programs, and activities presented in OMB reports. When we refer to respondents in this report, we mean federal officials who completed the questionnaire. Stakeholders refers to other individuals we interviewed who are experienced with federal funding for climate change programs and activities. HTML links to Web sites in this document are not maintained over time.

We conducted this performance audit from November 2009 to May 2011 in accordance with generally accepted government auditing standards. Those standards require that we plan and perform the audit to obtain sufficient, appropriate evidence to provide a reasonable basis for our findings and conclusions based on our audit objectives. We believe that the evidence obtained provides a reasonable basis for our findings and conclusions based on our audit objectives.

APPENDIX II. OMB REPORTS AND OTHER SOURCES OF CLIMATE CHANGE FUNDING INFORMATION

Table 1. Characteristics of Available OMB Funding Reports

Report year	Date issued	Legal requirement	Fiscal year budget authority data presented		
FY 2011	June 2010	Pub. L. No. 111-88, § 426 (2009)	2009 Actual	2010 Enacted	2011 Proposed
FY 2010	Not Applicable	None	None	None	None
FY 2009	Not Applicable	None	None	None	None
FY 2008	May 3, 2007	Pub. L. No. 110-5, § 104 (2007) (which continued in effect Pub. L. No. 109-102, § 585(b) (2005))	2006 Actual	2007 Enacted	2008 Proposed
FY 2007	April 2006	Pub. L. No. 109-102, § 585(b) (2005)	2005 Actual	2006 Enacted	2007 Proposed
FY 2006	March 2005	Pub. L. No. 108-447, § 576(b) (2004)	2004 Actual	2005 Enacted	2006 Proposed
FY 2005	May 2004	Pub. L. No. 108-199, § 555(b) (2004)	2003Actual	2004 Enacted	2005 Proposed
FY 2004	August 2003	Pub. L. No. 108-7, § 555(b) (2003)	2002 Actual	2003 Enacted	2004 Proposed
FY 2003	July 2002	Pub. L. No. 107-115, § 559(b) (2002)	2001 Actual	2002 Estimate	2003 Proposed

Source: GAO analysis of OMB reports.

Note: To view the full OMB reports, click on the report year in the table. OMB did not publish reports for fiscal years 2009 and 2010.

USGCRP program also reports funding for climate change science programs and activities in its annual reports to Congress.[64] In addition, both the Congressional Budget Office and Congressional Research Service recently reported on climate change funding.[65]

Congress periodically included provisions in appropriations laws requiring OMB to report funding for climate change programs and activities. As a result, OMB published several Federal Climate Change Expenditures Reports to Congress. Table 1 provides links to reports available on OMB's Web site, along with selected information about each report.

APPENDIX III. CLIMATE CHANGE FUNDING BY CATEGORY AS REPORTED BY OMB, 1993-2010

Budget authority in millions of dollars																		
Funding Category	**1993**	**1994**	**1995**	**1996**	**1997**	**1998**	**1999**	**2000**	**2001**	**2002**	**2003**	**2004**	**2005**	**2006**	**2007a**	**2008a**	**2009**	**2010**
Technology	$845	$1,038	$1,283	$1,106	$1,056	$1,251	$1,694	$1,793	$1,675	$1,637	$2,555	$2,868	$2,808	$2,78	$3,485	$4,196	$5,386	$5,504
Science	1,306	1,444	1,760	1,654	1,656	1,677	1,657	1,68	1,728	1,667	1,766	1,976	1,864	1,691	1,825	1,83	2,03	2,122
International assistance	201	186	228	192	164	186	325	177	218	224	270	252	234	249	188	227	373	1080
Wildlife adaptation					65													
Total	**$2,352**	**$2,668**	**$3,271**	**$2,952**	**$2,876**	**$3,114**	**$3,535**	**$3,511**	**$3,603**	**$3,522**	**$4,584**	**$5,090**	**$5,269**	**$5,876**	**$5,498**	**$6,255**	**$7,78**	**$8,771**

Source: GAO analysis of OMB reports

Notes: Blank cells indicate that OMB did not report a value for the account for that year. Totals may not add due to rounding.

Funding from the American Recovery and Reinvestment Act of 2009 (Pub. L. No. 111-5 (2009)) and climate change tax expenditures are not included in this table to maintain comparability with our 2005 report. See appendix V for more information about the Recovery Act and appendix VI for more information about tax expenditures.

a OMB did not publicly report climate change funding for these years because the annual appropriatio ns laws did not include a reporting requirement. OMB provided data for these years directly to GAO.

b Wildlife adaptation funding is an interim category while the Administration develops criteria to systematically account for a broader suite of adaptation programs.

Appendix IV. Climate Change Funding by Agency as Reported by OMB, 1993-2010

Budget authority in millions of dollars

Agency	1993	1994	1995	1996	1997	1998	1999	2000	2001	2002	2003	2004	2005	2006	2007[a]	2008[a]	2009	2010
Department of Energy	$963	$1,113	$1,173	$1,008	$968	$1,186	$1,536	$1,652	$1,665	$1,636	$2,214	$2,519	$2,469	$2,504	$3,158	$3,791	$4,711	$4,564
National Aeronautics and Space Administration	888	999	1,305	1,218	1,218	1,210	1,155	1,161	1,176	1,090	1,299	1,548	1,449	1,082	1,223	1,221	1,205	1,195
Department of Agriculture	55	56	60	52	57	53	138	132	54	59	104	116	110	110	109	116	322	567
Department of the Treasury	0	12	35	14	14	18	60	14	54	43	56	52	44	46	46	46	46	421
U.S. Agency for International Development	200	173	192	175	147	163	236	156	157	179	214	195	183	190	114	136	222	383
Department of Commerce	66	63	120	113	102	89	93	91	93	100	156	144	146	253	258	286	392	378
National Science Foundation	124	142	222	216	222	214	222	229	181	189	212	226	209	215	230	229	293	348

Budget authority in millions of dollars																	
Agency	**1993**	**1994**	**1995**	**1996**	**1997**	**1998**	**1999**	**2000**	**2001**	**2002**	**2003**	**2004**	**2005**	**2006**	**2007[a]**	**2008[a]**	**2009**	**2010**
Department of Defense											83	51	59	77	101	176	261	226
Department of State	1	1	1	3	3	5	7	7	7	7	6	6	7	12	41	59	55	199
Environm-ental Protection Agency	26	73	124	114	99	103	126	124	146	136	124	127	130	128	121	131	139	164
Department of Interior	22	29	27	26	26	26	27	27	27	26	29	29	29	27	27	34	45	128
Department of Transpo-rtation			5	6	13	5	3				27	9	3	17	18	20	45	128
Trade and Develop-ment Agency							16										10	17
Smithsoni-an Institution	7	7	7	7	7	7	7	7	7	6	6	6	6	6	6	6	6	7
Department of Health and Human Services						35	40	47	54	56	61	62	57	50	47	4	5	4
Millennium Challenge Corporati-on Department of Housing																		2

(Continued)

Budget authority in millions of dollars																		
Agency	**1993**	**1994**	**1995**	**1996**	**1997**	**1998**	**1999**	**2000**	**2001**	**2002**	**2003**	**2004**	**2005**	**2006**	**2007[a]**	**2008[a]**	**2009**	**2010**
and Urban Develop-ment							10	10										
Total	$2,352	$2,668	$3,271	$2,952	$2,876	$3,114	$3,535	$3,511	$3,603	$3,522	$4,584	$5,090	$4,900	$4,716	$5,499	$6,255	$7,757	$8,731[b]

Source: GAO analysis of OMB reports.

Notes: Blank cells indicate that OMB did not report a value for the account for that year. Totals may not add due to rounding.

[a] OMB did not publicly report climate change funding for these years because the annual appropriations laws did not include a reporting requirement. OMB provided data for these years directly to GAO.

[b] Total 2010 funding of $8.731 billion as presented by OMB and reported in this appendix does not match the total of $8.771 billion presented in appendixes III and V. In its June 2010 report that presented these data, OMB noted that totals may not add due to rounding and subtraction of double-counts.

Appendix V. Analysis of OMB Funding Report Accounts, 1993-2010

Budget authority in millions of dollars																			
ACCOUNT	**1993**	**1994**	**1995**	**1996**	**1997**	**1998**	**1999**	**2000**	**2001**	**200 2**	**2003**	**2004**	**2005**	**2006**	**2007[a]**	**2008[a]**	**2009**	**2010**	**ARR [b]**
TECHNOLOGY																			
Direct Technology																			
Department of Agriculture								0	03	3	42	45	8	4 49	48	51	271	453	0
Agricultural Research Service							0		0		2	2	2	2	2	2	5	5	0
Rural Business Service																			
Renewable energy program										0	22	23	23	23	23				
Value Added Producer Grants [c]													2	3	3	5	6	4	0
Rural Energy For America[c]																36	5	39	0
Biorefinery Assist-ance Programc																0	0	0	0
Bioenergy Prog-ram for Advanced Biofuelsc																	55	55	0
Forest Service																			
Forest and Range-land Research							0	0	3		1	0	2	2	2	1	1	5	0
Research and Development - Inventories of Carbon Biomass											1	0	1	1	1				

(Continued)

Budget authority in millions of dollars																			
ACCOUNT	**1993**	**1994**	**1995**	**1996**	**1997**	**1998**	**1999**	**2000**	**2001**	**2002**	**2003**	**2004**	**2005**	**2006**	**2007**[a]	**2008**[a]	**2009**	**2010**	**ARR**[b]
Natural Resources Conservation Service							0												
Carbon Cycle											1	1	1	1	1	0	9	7	0
Biomass Research and Development										3	14	14	13	14	14	0			
Cooperative State Research, Education and Extension Service																			
Biofuels/Biomass research; Formula Funds, National Research Initiative											3	5	5	3	3	5	5	5	0
Office of the Chief Economist - Methane to Markets [c]														0	0				
National Agricultural Statistics Service [c]																	0	0	0
Rural Business Service																			
Rural Energy For America [c, d]																0	55	60	0
Repowering Assistance Program [c, d]																0	35	0	0

Budget authority in millions of dollars																			
ACCOUNT	**1993**	**1994**	**1995**	**1996**	**1997**	**1998**	**1999**	**2000**	**2001**	**2002**	**2003**	**2004**	**2005**	**2006**	**2007[a]**	**2008[a]**	**2009**	**2010**	**ARR[b]**
Biorefinery Assistance Program [c, d]																0	75	245	0
National Institute of Food and Agriculture																			
Biomass Research and Development[c,d]																2	20	28	0
Department of Commerce													26	18	22	14	15	1 8	4
National Institutes of Standards and Technology											40	28	26	18					
Industrial Technical Services Advanced Technology Program											30	18	8	10	6	4	7	15	4
Scientific and Technical Research Services							0	2			10	10	18	8					
International Trade Administration																			
Operations and Administration [c]													0	0	0	2	2	2	0
Department of Defense											83	51	59	77	101	176	261	226	139
Research, Development, Test and Evaluation, Army											45	15	27	49	69	98	98	93	44

(Continued)

Budget authority in millions of dollars																		
ACCOUNT	**1993**	**1994**	**1995**	**1996**	**1997**	**1998**	**1999**	**2000**	**2001**	**2002**	**2003**	**2004**	**2005**	**2006**	**2007[a]**	**2008[a]**	**2009**	**2010**	**ARR[b]**
Research, Development, Test and Evaluation, Navy											16	17	18	17	13	44	54	13	18
Research, Development, Test and Evaluation, Air Force											3	1	1	0	13	34	108	120	35
Research, Development, Test and Evaluation, Defense-wide											19	19	13	11	6	0	0	0	42
Defense Advanced Research Projects Agency [c]												17	11	7	6				
Office of the Secretary of Defense [c]												2	2	4	0				
Department of Energy	595	753	829	683	658	729	890	980	1,050	1,519	2,099	2,390	2,342	2,374	3,032	3,663	4,543	4,399	25,223
Energy Conservation										897	880	868							
Energy Conservation Research and Development	346	435	468	415	414	457	518	577	619	622									
State Energy Grants										45									
Weatherization										230									

Budget authority in millions of dollars																			
ACCOUNT	**1993**	**1994**	**1995**	**1996**	**1997**	**1998**	**1999**	**2000**	**2001**	**2002**	**2003**	**2004**	**2005**	**2006**	**2007[a]**	**2008[a]**	**2009**	**2010**	**ARR[b]**
Energy Supply / Energy Supply and Conservation	249	318	361	268	244	272	332	315	375	400	667								
Nuclear Energy Research Initiative										32									
Electricity/ Electricity Supply and Distribution / Electricity Delivery and Energy Reliability											88	73	57	77	120	130	113	109	4340
Renewables	249	318	361	268	244	272	332	310	370	368	322	352							
Nuclear							0	5	5		257	309	291	343	513	682	787	747	0
Energy Efficiency and Renewable Energy [c]													1,234	1,166	1,411	1,722	2,179	2,242	16,800
Fossil Energy Research and Development							24	52	18	184	253	455	374	397	493	611	762	560	3,400
Sequestration Research and Development										32									
Greenhouse Gas Emission Reduction										152									
Science							13	33	35	35	298	333	386	391	487	512	700	723	283
Sequestration										32									
Energy Information Administration							3	3	3	3									

(Continued)

Budget authority in millions of dollars																		
ACCOUNT	**1993**	**1994**	**1995**	**1996**	**1997**	**1998**	**1999**	**2000**	**2001**	**2002**	**2003**	**2004**	**2005**	**2006**	**2007[a]**	**2008[a]**	**2009**	**2010**	**ARR[b]**
Departmental Administration – Climate Change Technology Prog-ram Direction[c]												0	0	0	1	1	2	9	0
Innovative Techn-ology Loan Guara-ntee Program [c]														0	7	5			
Advanced Research Projects Agency																			
Energy (ARPA-E)[c]																	0	9	0
Energy (ARPA-E) Recovery Act [c]																0	0	0	400
Environmental Protection Agency		43	102	96	86	90	109	103	123	115	102	110	110	109	105	114	111	133	0
Environmental Programs and Management		35	91	81	70	73	72	76	96	89	82	89	91	90	92	97	94	113	0
Science and Technology		8	11	15	16	17	37	27	27	26	20	22	19	19	13	17	17	20	0
Department of Housing and Urban Development																			

Budget authority in millions of dollars																			
ACCOUNT	**1993**	**1994**	**1995**	**1996**	**1997**	**1998**	**1999**	**2000**	**2001**	**2002**	**2003**	**2004**	**2005**	**2006**	**2007[a]**	**2008[a]**	**2009**	**2010**	**ARR[b]**
Research and Technology (PATH)							10	10											
Department of Interior											1	1	2						
U.S. Geological Survey - Surveys, Investigations and Research																			
Geology Discipline, Energy Program											1	1	2						
National Aeronautics and Space Administration																			
Exploration, Science & Aeronautics											152	227	208	129	139	137	119	124	31
National Science Foundation																			
Research and Related Activities											9	11	11	18	21	22	24	26	2
Department of Transportation											27	5	2	16	17	19	43	125	100
Federal Transit Administration																			
Capital Investment Grants											26								
Research and University Research Centers and														14	16	18	19	94	100

(Continued)

Budget authority in millions of dollars																		
ACCOUNT	**1993**	**1994**	**1995**	**1996**	**1997**	**1998**	**1999**	**2000**	**2001**	**2002**	**2003**	**2004**	**2005**	**2006**	**2007[a]**	**2008[a]**	**2009**	**2010**	**ARR[b]**
Formula and Bus Grants[c]																			
Office of the Secretary of Technology																			
Transportation, Policy, Research and Development											1	4	1						
Research and Special Programs Administration																			
Research and Special Programs											0								
National Highway Traffic Safety Administration												0	[e]	1	1	1	0	1	0
Research and Innovative Technology Administration																			
Research and Development												1	1	1	1	1	1	1	0
Federal Highways Administration																			
Federal-aid Highways[c]																	19	19	0
Federal Aviation Administration																			

Budget authority in millions of dollars																		
ACCOUNT	**1993**	**1994**	**1995**	**1996**	**1997**	**1998**	**1999**	**2000**	**2001**	**2002**	**2003**	**2004**	**2005**	**2006**	**2007[a]**	**2008[a]**	**2009**	**2010**	**ARR[b]**
Research, Engineering, and Development [c]																	3	11	0
Direct Technology Total	$595	$796	$931	$779	$744	$819	$1,009	$1,095	$1,176	$1,637	$2,555	$2,868	$2,808	$2,789	$3,485	$4,196	$5,386	$5,504	$25,499
National Climate Change Technology Initiative [c]																			
Department of Energy													162	289					
Energy Supply and Conservation																			
Energy Efficiency and Renewable Energy [c]													65	66					
Nuclear [c]													9	102					
Fossil Energy Research and Development																			
Efficiency and Sequestration [c]													89	121					
Departmental Administration-Climate Change Technology Program Direction[c]													0	0					
Environmental Protection Agency																			

(Continued)

Budget authority in millions of dollars																			
ACCOUNT	**1993**	**1994**	**1995**	**1996**	**1997**	**1998**	**1999**	**2000**	**2001**	**2002**	**2003**	**2004**	**2005**	**2006**	**2007[a]**	**2008[a]**	**2009**	**2010**	**ARR[b]**
Environmental Programs and Management [c]													11	10					
National Climate Change Techn-ology Initiative Total													$173	$299					
Indirect Technology																			
Department of Energy	250	242	231	212	201	351	417	434	499										
Fossil Energy Research and Development	250	242	231	212	201	196	233	243	274										
Coal-efficient combustion & utilization	186	166	144	120	101	105													
Natural gas—efficient combus-tion & utilization	64	76	87	92	100	91													
Energy Supply																			
Nuclear Energy Research and Development						0	18	22	34										
Energy Conser-vation Research and Development																			

Budget authority in millions of dollars

ACCOUNT	1993	1994	1995	1996	1997	1998	1999	2000	2001	2002	2003	2004	2005	2006	2007[a]	2008[a]	2009	2010	ARR[b]
Weatherization &						155	166	169	191										
State Energy Grants																			
Department of Agriculture							86	76											
Agriculture Research Service							44	46											
Cooperative State Research, Education, & Extension Service																			
Research and Education Assistance							11	11											
Initiative for Future Agriculture & Food Systems								9											
Forest Service																			
Forest and Rangeland Management							9	9											
Executive Operations							1	1											
Departmental Administration							[e]	[e]											
Alternative Agricultural Research and Commercialization							4												
Natural Resources Conservation							16												

(Continued)

Budget authority in millions of dollars																		
ACCOUNT	**1993**	**1994**	**1995**	**1996**	**1997**	**1998**	**1999**	**2000**	**2001**	**2002**	**2003**	**2004**	**2005**	**2006**	**2007[a]**	**2008[a]**	**2009**	**2010**	**ARR[b]**
Service - Forestry Incentives Program																			
Rural Development - Rural Community Advancement Program							1												
Department of Energy							109		124										
Energy Supply																			
Solar and Renewable Energy Research and Development							40	70											
Energy Conservation Research and Development							41	11											
Fossil Energy Research and Development							0	13											
Science (Basic Science)							27	30											
Partnership for a New Generation of Vehicles							73	64											

Budget authority in millions of dollars																		
ACCOUNT	**1993**	**1994**	**1995**	**1996**	**1997**	**1998**	**1999**	**2000**	**2001**	**2002**	**2003**	**2004**	**2005**	**2006**	**2007**[a]	**2008**[a]	**2009**	**2010**	**ARR**[b]
Department of Commerce			63	56	42	29	30	22											
Under Secretary for Technology/ Office of Technology Policy							1	0											
Salaries and Expenses			0	1	1	1													
National Institutes of Standards and Technology							29	22											
Scientific and Technical Resea-rch and Services			7	7	7	6													
Industrial Technology Services			56	48	34	22													
National Science Foundation																			
Research and Related Activities			53	53	56	47	40	42											
Department of Transportation																			
National Highway Traffic Safety Administration (and FTA prior to FY 1999)																			
Operations and Research			5	6	13	5	3												

(Continued)

Budget authority in millions of dollars																			
ACCOUNT	**1993**	**1994**	**1995**	**1996**	**1997**	**1998**	**1999**	**2000**	**2001**	**2002**	**2003**	**2004**	**2005**	**2006**	**2007[a]**	**2008[a]**	**2009**	**2010**	**ARR[b]**
Indirect Technology Total	$250	$242	$352	$327	$312	$432	$685	$698	$499	[f]	[f]	[f]	[f]	[f]	[f]	[f]	[f]	[f]	[f]
Technology Total	$845	$1,038	$1,283	$1,106	$1,056	$1,251	$1,694	$1,793	$1,675	$1,637	$2,555	$2,868	$2,808	$2,789	$3,485	$4,196	$5,386	$5,504	$25,499
SCIENCE																			
U.S. Global Change Research Program																			
Department of Agriculture	55	56	60	52	57	53	52	56	51	56	60	70	62	61	61	65	47	109	0
Agricultural Research Service	17	18	24	24	26	27	26	28	29	30	35	37	38	38	40	39	20	20	0
Cooperative State Research, Education, & Extension Services																			
Research and Education	11	12	10	10	12	7	7	9	4	9	8	16	5	4	2	4			
Economic Research Service	1	1	1	1	1	1	1	1	1	e	0	0	0	0	0	0	0	1	0
Natural Resources Conservation Service																			
Conservation Operations	2	2	2	2	1	1	1	1											
Forest Service																			
Forest and Range-land Research	24	23	23	15	17	17	17	17	17	17	17	17	18	18	19	22	22	32	0

Budget authority in millions of dollars																		
ACCOUNT	**1993**	**1994**	**1995**	**1996**	**1997**	**1998**	**1999**	**2000**	**2001**	**2002**	**2003**	**2004**	**2005**	**2006**	**2007[a]**	**2008[a]**	**2009**	**2010**	**ARR[b]**
National Institute of Food and Agriculture [c]																	5	56	0
Department of Commerce														235	236	272	377	360	218
National Oceanic and Atmospheric Administration																			
Operations, Research, and Facilities	66	63	57	57	60	60	63	67	93	100	98	116	120	226	229	265	274	309	0
Procurement, Acquisition and Construction[c]														9	7	7	101	49	218
National Institute of Standards and Technology															0	0	2	2	0
Department of Energy																			
Science (Biological & Environmental Research)	118	118	113	113	109	106	114	114	116	117	112	129	127	130	126	128	168	165	65
Environmental Protection Agency																			
Science and Technology	26	30	22	18	13	13	17	21	23	21	22	17	20	19	16	17	18	21	0
Department of Health and Human Services						35													
National Institutes of Health						35	40	47	54	56	61	62	57	50	47	4	5	4	0

(Continued)

Budget authority in millions of dollars																		
ACCOUNT	**1993**	**1994**	**1995**	**1996**	**1997**	**1998**	**1999**	**2000**	**2001**	**2002**	**2003**	**2004**	**2005**	**2006**	**2007[a]**	**2008[a]**	**2009**	**2010**	**ARR[b]**
National Institute of Environmental Health Sciences						4													
National Eye Institutes						9													
National Cancer Institute						21													
National Institute of Arthritis and Musculoskeletal & Skin Diseases						[e]													
Department of the Interior																			
U.S. Geological Survey																			
Surveys, Investigations, and Research	22	29	27	26	26	26	27	27	27	26	2 8	28	27	27	27	34	45	63	0
National Aeronautics and Space Administration															1,084	1,084	1,086	1,071	237
Science, Aeronautics, and Technology	888	999	1305	1218	1218	1210	1155	1161	1176	1090	1144	1321	1241	953	1,084	1,084			
Science [c]															0	0	1,086	1,071	237

Budget authority in millions of dollars																			
ACCOUNT	**1993**	**1994**	**1995**	**1996**	**1997**	**1998**	**1999**	**2000**	**2001**	**2002**	**2003**	**2004**	**2005**	**2006**	**2007[a]**	**2008[a]**	**2009**	**2010**	**ARR[b]**
National Science Foundation																			
Research and Related Activities	124	142	169	163	166	167	182	187	181	189	188	215	198	197	207	207	269	319	121
Smithsonian Institution																			
Salaries and Expenses	7	7	7	7	7	7	7	7	7	6	6	6	6	6	6	6	6	7	0
Department of State																			
International Organizations and Programs[c]												1							
Department of Transportation												4	1	1	1	1	2	3	0
Federal Highways Administration																			
Federal-aid Highways [c]												4	1	[e]	[e]	1	[e]	0	0
Federal Aviation Administration																			
Research, Engineering, and Development [c]												0	[e]	[e]	[e]	[e]	2	3	0
Federal Transit Administration																			
Research and University Research Centers [c]														[e]	[e]	[e]	[e]	[e]	0

(Continued)

Budget authority in millions of dollars																			
ACCOUNT	**1993**	**1994**	**1995**	**1996**	**1997**	**1998**	**1999**	**2000**	**2001**	**2002**	**2003**	**2004**	**2005**	**2006**	**2007[a]**	**2008[a]**	**2009**	**2010**	**ARR[b]**
and Formula and Bus Grants																			
U.S. Agency for International Development																			
Development Assistance										6	6	6	6	13	14	14	17	36	0
U.S. Global Change Research Program Total	$1,306	$1,444	$1,760	$1,654	$1,656	$1,677	$1,657	$1,687	$1,728	$1,667	$1,725	$1,803	$1,660	$1,488	$1,825	$1,832	$2,023	$2,122	$641
Climate Change Research Initiative																			
Department of Agriculture											2	6	8	8					
Agricultural Research Service											0	1	2	2					
Forest Service																			
Forest and Range-land Research											1	5	6	6					
Department of Commerce																			
National Oceanic and Atmospheric Administration																			
Operations, Rese-arch, and Facilities											18	34	46	34					

Budget authority in millions of dollars																		
ACCOUNT	**1993**	**1994**	**1995**	**1996**	**1997**	**1998**	**1999**	**2000**	**2001**	**2002**	**2003**	**2004**	**2005**	**2006**	**2007[a]**	**2008[a]**	**2009**	**2010**	**ARR[b]**
Department of Energy																			
Science (Biological & Environmental Research)											3	27	25	28					
National Aeronautics and Space Administration																			
Science, Aeronautics, and Technology											3	65	94	95					
National Science Foundation																			
Research and Related Activities											15	30	25	25					
Department of State																			
International Organizations and Programs												1							
Department of Transportation												4		1					
Federal Highway Administration																			
Federal Aid - Highways												4	1	0					
Federal Transit Administration																			

(Continued)

Budget authority in millions of dollars																		
ACCOUNT	**1993**	**1994**	**1995**	**1996**	**1997**	**1998**	**1999**	**2000**	**2001**	**2002**	**2003**	**2004**	**2005**	**2006**	**2007[a]**	**2008[a]**	**2009**	**2010**	**ARR[b]**
Formula Grants and Researchc																			
Federal Aviation Administration																			
Research, Engineering, and Developmentc												0	0						
U.S. Agency for International Development																			
Development Assistance												6	6	13					
Climate Change Research Initiative Total											$41	$173	$204	$203					
Science Total	$1,306	$1,444	$1,760	$1,654	$1,656	$1,677	$1,657	$1,687	$1,728	$1,66 7	$1,76 6	$1,97 6	$1,864	$1,691	$1,825	$1,832	$2,023	$2,122	$641
INTERNATIO-NAL ASSISTANCE Core Agencies																			
Department of State												5	7	12	41	59	55	199	
International Organizations and Programs	1	1	1	3	3	5	7	7	7	7	6	5	6	6	6	29	29	39	

Budget authority in millions of dollars																			
ACCOUNT	**1993**	**1994**	**1995**	**1996**	**1997**	**1998**	**1999**	**2000**	**2001**	**2002**	**2003**	**2004**	**2005**	**2006**	**2007[a]**	**2008[a]**	**2009**	**2010**	**ARR[b]**
Economic Support Fund[c]												0	1	6	32	26	24	158	
Diplomatic and Consular Affairs[c]													0	0	3	4	2	2	
Department of the Treasury										43	56	52	44	46	46	46	46	421	
International Development Assistance																			
Global Environment Facility[g]		12	35	14	14	18	60	14	41	38	56	32	20	26	26	26	26	26	
Debt Restructuring																			
Tropical Forest Conservation									3	1 5		20	24	20	20	20	20	20	
Asian Development Bank[c]													0						
Clean Technology Fund[c]															0	0	0	300	
Strategic Climate Fund[c]																0	0	75	
U.S. Agency for International Development							236	156	157	174	208	195	183	190	100	122	222	383	
Development Assistance (DA)	200	173	192	175	147	163	169	109	112	116	140	125	134	118	89	88	113	313	
Development Credit Authority (DCA)							1	1	1										

(Continued)

Budget authority in millions of dollars																		
ACCOUNT	**1993**	**1994**	**1995**	**1996**	**1997**	**1998**	**1999**	**2000**	**2001**	**2002**	**2003**	**2004**	**2005**	**2006**	**2007[a]**	**2008[a]**	**2009**	**2010**	**ARR[b]**
Economic Support Fund							19	8		12	6	9	5	33	0	6	94	44	
Assistance for the Independent States of the Former Sov-iet Union (FSA)							35	34	31	30	48	47	34	30	5	15			
Assistance for Eastern Europe and the Baltic States (AEEB)								12	43	111	8	7	5	6	3	11			
Assistance for Europe, Eurasia, and Central Asia[c]																0	15	26	
International Disaster Assistance (IDA)										4	4	2	2	2	2	2	0	0	
Andean Counterd-rug Initiative(ACI)											2	3	2	0	0				
Pub. L. 480 Title II Food Aid [c]												1	1	1	0				
Core Agencies Total																	**$323**	**$1,003**	
Complementary Agenciesc																			
Department of Agriculture																			
Forest Service																			

Budget authority in millions of dollars																			
ACCOUNT	**1993**	**1994**	**1995**	**1996**	**1997**	**1998**	**1999**	**2000**	**2001**	**2002**	**2003**	**2004**	**2005**	**2006**	**2007[a]**	**2008[a]**	**2009**	**2010**	**ARR[b]**
Forest and Rangeland Research [c]																	4	5	
Department of Commerce																	11	11	
National Oceanic and Atmospheric Administration																			
Operations, Resea-rch and Facilities [c]																	9	9	
International Trade Administration																			
Operations and Administration [c]																	2	2	
Department of Energy																	0	13	
Energy Supply																			
Solar and Renewable Energy Research and Development										6									
Energy Efficiency and Renewable Energy [c]																	0	8	
Fossil Energy Research and Development																			
Efficiency and Sequestration [c]																	0	3	

(Continued)

Budget authority in millions of dollars																		
ACCOUNT	**1993**	**1994**	**1995**	**1996**	**1997**	**1998**	**1999**	**2000**	**2001**	**2002**	**2003**	**2004**	**2005**	**2006**	**2007[a]**	**2008[a]**	**2009**	**2010**	**ARR[b]**
Environmental Protection Agency																			
Science [c]																	0	3	
Environmental Programs and Management [c]																	20	21	
Millennium Challenge Corporation[c]																	0	2	
National Aeronautics and Space Administration																			
Science [c]																	2	2	
National Science Foundation																			
Research and Related Activities [c]																	3	6	
US Trade and Development Agency							16										10	17	
Complementary Agencies Total																	$50	$77	
International Assistance Total	**$201**	**$186**	**$228**	**$192**	**$164**	**$186**	**$325**	**$177**	**$218**	**$224**	**$270**	**$252**	**$234**	**$249**	**$188**	**$227**	**$373**	**$1,080**	
WILDLIFE ADAPTATION[h]																			

Budget authority in millions of dollars																			
ACCOUNT	**1993**	**1994**	**1995**	**1996**	**1997**	**1998**	**1999**	**2000**	**2001**	**2002**	**2003**	**2004**	**2005**	**2006**	**2007[a]**	**2008[a]**	**2009**	**2010**	**ARR[b]**
Department of Interior																			
National Park Service																			
Operation of the National Park Service[c]																	0	10	0
Fish and Wildlife Service																			
Resource Managementc																	0	40	0
Bureau of Land Management -																			
Management of Lands and Resourcesc																	0	15	0
WILDLIFE ADAPTATION TOTAL																	0	$65	0
Total Climate Change Funding	**$2,352**	**$2,668**	**$3,271**	**$2,952**	**$2,876**	**$3,114**	**$3,535**	**$3,511**	**$3,603**	**$3,522**	**$4,584**	**$5,090**	**$5,269**	**$5,876**	**$5,498**	**$6,255**	**$7,782**	**$8,771**	**$26,140**

Source: GAO analysis of OMB reports.

Notes: GAO calculated the total for shaded cells based on OMB data presented in its reports.

Blank cells indicate that OMB did not report a value for the account for that year.

Totals may not add due to rounding.

Italics indicate that the number does not add to the section total because it is already counted elsewhere in the table.

Climate-related tax expenditures are not included in this table to maintain comparability with our 2005 report, *Climate Change: Federal Reports on Climate Change Funding Should Be Clearer and more Complete,* GAO-05-461. Climate-related tax expenditures are presented separately in appendix VI.

[a] OMB did not publicly report climate change funding for these years because the annual appropriations laws did not include a reporting requirement. OMB provided data for these years directly to GAO.

[b] In its June 2010 report, OMB reported funding for climate change programs and activities in the American Recovery and Reinvestment Act of 2009 (Pub. L. No. 111-5 (2009)).

[c] Funding for this account has been added since our 2005 report, *Climate Change: Federal Reports on Climate Change Funding Should Be Clearer and more Complete,* GAO-05-461.

[d] OMB identified funding for this account as mandatory under the Food, Conservation, and Energy Act of 2008 (Pub. L. No. 110- 234 (2008)).

[e] OMB presented funding of less than $500,000 for this account.

[f] OMB did not distinguish between indirect and direct technology funding for this year.

[g] GEF funding as presented by OMB for each year represents the portion of total GEF funding that is related to climate change.

[h] Wildlife adaptation funding is an interim category while the Administration develops criteria to systematically account for a broader suite of adaptation programs.

APPENDIX VI. CLIMATE CHANGE TAX EXPENDITURES AS REPORTED BY OMB, 2003-2010

Table 2. Estimated Revenue Loss from Energy Tax Provisions That May Reduce Greenhouse Gases as Reported by OMB, 2003-2010

Revenue effect in millions of dollars								
	2003	**2004**	**2005**	**2006**	**2007[a]**	**2008[a]**	**2009**	**2010**
Tax provisions that may reduce greenhouse gases								
New technology credit (without coal)[b]	$380	$330	$219	$440	$410	$900	360 $	$780
Credit and deduction for clean-fuel burning vehicles	90	70	70	110	260	170	130	240
Exclusion of utility conservation subsidies	110	100	80	110	120	120	140	140
Credit for holding clean renewable energy bonds			0	20	20	40	70	80
Allowance of deduction for certain energy efficient commercial building property			0	80	190	170	60	80
Credit for construction of new energy efficient homes			0	10	20	30	30	20
Credit for energy efficiency improvements to existing homes			0	230	380	230	570	1950
Credit for energy efficient appliances			0	120	80	120	130	130
Credit for residential purchases / installations of solar and fuel cells			0	10	10	20	110	80
Credit for business installation of qualified fuel cells			0	30	30	0		
Energy Investment Credit[c]						40	270	530
Qualified energy conservation bonds						0	0	10
Total energy tax provisions that may reduce greenhouse gases	$580	$500	$369	$1,160	$1,520	$1,840	$1,870	$4,140

Table 2. (Continued)

Revenue effect in millions of dollars								
	2003	**2004**	**2005**	**2006**	**2007[a]**	**2008[a]**	**2009**	**2010**
Energy grants								
Revenue effect in millions of dollars								
	2003	**2004**	**2005**	**2006**	**2007[a]**	**2008[a]**	**2009**	**2010**
Energy Grants in Lieu of New Technology Credit or Energy Investment Credit[d]							1,050	3,090
Total (tax provisions plus grants)	**$580**	**$500**	**$369**	**$1,160**	**$1,520**	**$1,840**	**$2,920**	**$7,230**

Source: GAO analysis of OMB reports.

Notes: OMB did not report revenue effects for existing tax expenditures that may reduce greenhouse gases from 1993 through 2002. OMB began reporting revenue effects for existing climate-related tax expenditures in response to recommendations from our 2005 report, *Climate Change: Federal Reports on Climate Change Funding Should Be Clearer and more Complete,* GAO-05-461. Data for 2003 and 2004 were presented in OMB's April 2006 report.

Blank cells indicate that OMB did not report a value for the account for that year.

[a] OMB did not publicly report climate change funding for these years because the annual appropriations laws did not include a reporting requirement. OMB provided data for these years directly to GAO.

[b] Estimates of revenue loss from coal provisions have been removed from the tax expenditure estimate in the budget. In previous years, the Energy Investment Credit was contained within the New Technology Credit.

[in] previous years, the Energy Investment Credit was contained within the New Technology Credit. The Energy Investment Credit also includes the business installation of fuel cells, which was an independent entry in tables from previous years. Estimates of revenue loss from the micro-turbine provision have been removed from the tax expenditure estimate in the budget.

[d] Firms can take an energy grant in lieu of the energy production credit or the energy investment credit for facilities placed in service in 2009 and 2010 or whose construction commenced in 2009 and 2010. The grants are considered outlays and are direct substitutes for the energy tax provisions.

Appendix VII. Source Documents Articulating Federal Strategic Climate Change Priorities

Our review of questionnaire responses, available literature, and interviews with stakeholders found that federal climate change priorities are presented in six general sources, including (1) strategic plans for interagency programs and agencies, (2) executive level guidance memoranda, (3) the development of new interagency initiatives, (4) regulations and guidance memoranda, and (5) international commitments, and (6) testimony of federal executives before Congress.[66]

Strategic Plans

Two interagency programs that coordinate federal climate change activities have strategic plans with explicit goals. In its 2003 strategic plan, USGCRP identifies five strategic goals to focus and orient research in the program: (1) improve knowledge of the Earth's past and present climate and environment, including its natural variability, and improve understanding of the causes of observed variability and change; (2) improve quantification of the forces bringing about changes in the Earth's climate and related systems; (3) reduce uncertainty in projections of how the Earth's climate and related systems may change in the future; (4) understand the sensitivity and adaptability of different natural and managed ecosystems and human systems to climate and related global changes; and (5) explore the uses and identify the limits of evolving knowledge to manage risks and opportunities related to climate variability and change.[67] OMB's June 2010 report to Congress on climate change funding provides a slightly different interpretation of the USGCRP strategic plan, with an increased emphasis on providing useable information for decision makers.[68]

The USGCRP strategic plan includes a range of approaches to work toward these goals. The plan also spells out criteria for establishing funding priorities. For example, the USGCRP strategic plan states that programs with good track records of past performance will be favored for continued investment to the extent that timetables and metrics for evaluating future progress are provided. USGCRP's strategic plan was revised in 2008, but the overall goals were not altered.[69] USGCRP has kicked off a strategic planning process that will yield a new research plan in 2011 and a full strategic plan in 2013, according to OSTP.

USGCRP utilizes reports and discussions conducted by the National Academies as a source of input to its planning.[70] According to OSTP, the National Academies has been very valuable in advising the U.S. government on strategic priorities as well as specific programmatic directions. Most recently, the *America's Climate Choices* suite of studies, which were supported by the National Oceanic and Atmospheric Administration (NOAA), has proven very helpful guiding the strategic direction for a variety of activities currently under way, according to OSTP.[71]

In its September 2006 strategic plan, CCTP sets six strategic goals to, in part, enable the stabilization of greenhouse gas concentrations at a level that would prevent dangerous anthropogenic interference.[72] These goals are (1) reduce emissions from energy end use and infrastructure; (2) reduce emissions from the energy supply; (3) capture and sequester carbon dioxide; (4) reduce emissions of noncarbon dioxide greenhouse gases; (5) improve capabilities to measure and monitor greenhouse gas emissions; and (6) bolster basic science contributions to technology development. CCTP's strategic plan also includes a list of core approaches and federal programs that contribute to achieving these goals. The plan also includes a prioritization process with planning principles and investment criteria to, for example, maximize expected return on investment. CCTP's strategic plan has not been revised, but a CCTP official stated that the priorities established in its 2006 strategic plan remain essentially unchanged, with the exception of an increased focus on geoengineering and adaptation.[73]

Individual agencies are also including climate change in their strategic plans. For example, U.S. Department of Agriculture's fiscal year 2010–2015 strategic plan includes a strategic goal to ensure national forests are conserved, restored, and made more resilient to climate change while enhancing water resources.[74] Similarly, sections of the U.S. Department of the Interior's proposed fiscal year 2011-2016 strategic plan relate to climate change adaptation activities. Further, within the Department of the Interior, the U.S. Fish and Wildlife Service developed a strategic plan for responding to climate change. Other agencies are undertaking similar efforts.

Executive Level Guidance Memoranda

Executive level guidance memoranda are a mechanism for OMB and OSTP to define climate change priorities within the overall federal budget. The Directors of OMB and OSTP described climate change priorities within the science and technology budget in a July 21, 2010, memorandum for the heads of executive departments and agencies titled *Science and Technology Priorities for the FY 2012 Budget.*[75] The memorandum instructs agencies to explain in their budget submissions how they will redirect available resources, as appropriate, from lower-priority areas to science and technology activities that address six challenges. One of the six challenges identified in the memorandum is "understanding, adapting to, and mitigating the impacts of global climate change." Specifically, agencies are requested to identify the activities in their budgets that support two priority areas—the National Assessment (described in more detail below) and the monitoring of greenhouse gas emissions. OMB and OSTP also issued supplemental guidance on climate change science collaboration on August 13, 2010.[76]

The Directors of OMB and OSTP issued a similar memorandum for the fiscal year 2011 budget on August 4, 2009.[77] One of the four key challenges identified in that memorandum was promoting innovative energy technologies to reduce dependence on energy imports and mitigate the impact of climate change while creating green jobs and new businesses.

Development of New Interagency Initiatives

Several new formal and informal interagency task forces and working groups managed by entities within the Executive Office of the President also demonstrate current climate change priorities. These include efforts to implement a 2009 Executive Order on federal sustainability, the Climate Change Adaptation Task Force, an Interagency Task Force on Carbon Capture and Storage, the creation of a recurring national climate assessment process, the development of a NOAA climate service, and other formal and informal efforts. According to OSTP, overarching climate-related policy directions are coordinated in part by the "Green Cabinet."

Federal Agency Strategic Sustainability Performance Plans

The October 5, 2009, Executive Order 13514 on Federal Leadership in Environmental, Energy, and Economic Performance set sustainability goals and targets, such as greenhouse gas emissions reductions and energy efficiency improvements, for federal agencies to meet.[78] The Executive Order requires all federal agencies to (1) annually submit a comprehensive inventory of greenhouse gas emissions and (2) prepare a strategic sustainability performance plan that includes, among other things, a greenhouse gas reduction target for fiscal year 2020 and several water consumption and waste reduction targets. Agencies are to integrate sustainability performance plans into their strategic planning and budget processes. CEQ released guidance on Federal Greenhouse Gas Reporting and Accounting on October 6, 2010. The guidance establishes methods for calculating and reporting greenhouse gas emissions associated with federal agency operations. The White House released federal agency Strategic Sustainability Performance Plans on September 9, 2010.[79]

Climate Change Adaptation Task Force

Executive Order 13514 also called for federal agencies to participate actively in the already existing Interagency Climate Change Adaptation Task Force.[80] The task force, which began meeting in Spring 2009, is co- chaired by CEQ, NOAA, and OSTP, and includes representatives from more than 20 federal agencies and executive branch offices. The task force was formed to develop federal recommendations for adapting to climate change impacts both domestically and internationally, and to recommend key components to include in a national strategy.

On October 14, 2010, the task force released its interagency report outlining recommendations to the President for how federal policies and programs can better prepare the United States to respond to the impacts of climate change. The report recommends that the federal government implement actions to expand and strengthen the nation's capacity to better understand, prepare for, and respond to climate change. These recommended actions include making adaptation a standard part of agency planning to ensure that resources are invested wisely and services and operations remain effective in a changing climate. According to CEQ, the task force will continue to meet as an interagency forum for discussing the federal government's adaptation approach and to support and monitor the implementation of recommended actions in the progress report. It will prepare another report in October 2011 that documents progress toward implementing its recommendations and provides additional recommendations for refining the federal approach to adaptation, as appropriate, according to CEQ.[81]

Related to the efforts of the Adaptation Task Force, OMB is beginning to account for federal climate change adaptation expenditures. In its June 2010 report to Congress, OMB summarized certain activities at the Department of the Interior designed to promote wildlife adaptation, and noted that it is working to develop criteria to systematically account for a broader suite of adaptation programs.[82]

Carbon Capture and Storage Task Force

The Interagency Task Force on Carbon Capture and Storage was established on February 3, 2010, to develop a comprehensive and coordinated federal strategy to speed the commercial development and deployment of clean coal technologies.[83] The task force, co-chaired by the Department of Energy and the Environmental Protection Agency, was charged with proposing a plan to overcome the barriers to the widespread, cost-effective deployment of carbon capture and storage within 10 years, with a goal of bringing 5 to 10 commercial demonstration projects online by 2016. The task force issued its report on August 12, 2010.[84] The report concludes that carbon capture and storage can play an important role in domestic greenhouse gas emissions reductions while preserving the option of using abundant domestic fossil energy resources. However, widespread cost-effective deployment of carbon capture and storage will occur only if the technology is commercially available at economically competitive prices and supportive national policy frameworks are in place. The task force's recommendations include specific actions to help overcome remaining barriers to deployment.

National Climate Assessment

The Global Change Research Act of 1990 mandates that USGCRP prepare an assessment periodically, but at least every 4 years, which analyzes the effects of global change on the natural environment and biological diversity, among other things.[85] According to an OSTP official, the next assessment mandated by the act will be released in 2013. To meet this deadline, OSTP is leading an interagency effort under the auspices of USGCRP to establish a recurring National Climate Assessment. According to an OSTP official, this effort is coordinated through the Interagency National Climate Assessment Task Force, which represents 18 agencies and departments.[86]

OSTP is working with agencies and USGCRP to develop the scope and plan for the assessment. According to a September 7, 2010 *Federal Register* notice, the National Climate Assessment to be released in 2013 is envisioned as a comprehensive assessment of climate change, impacts, vulnerabilities and response strategies within a context of how communities and the nation as a whole create sustainable and environmentally sound development paths.[87] The notice states that the primary vision of the National Climate Assessment is a continuing, inclusive national process that (1) synthesizes relevant science and information; (2) increases understanding of what is known and not known; (3) identifies needs for information related to preparing for climate variability and change and reducing climate impacts and vulnerability; (4) evaluates progress of adaptation and mitigation activities; (5) informs science priorities; (6) builds assessment capacity in regions and sectors; and (7) builds societal understanding and skilled use of assessment findings.

NOAA Climate Service

In an announcement on February 8, 2010, the Department of Commerce proposed establishing a NOAA climate service. Though not established, planning information is available on the NOAA climate service Web site, including draft vision and strategic framework documents.[88] According to NOAA, such a climate service would provide a single, reliable, and authoritative source for climate data, information, and decision-support services to help individuals, businesses, communities, and governments make smart choices in anticipation of a climate changed future. A NOAA climate service would provide a one-stop shop for users across the nation, according to NOAA, and would also bring together many of the agency's existing climate assets including research labs, climate observing systems, modeling facilities, and monitoring systems.[89]

Other Formal and Informal Interagency Task Forces and Working Groups

According to CEQ, OMB, and OSTP, there are other climate-related formal and informal interagency task forces and working groups within the federal government. These groups include

- *National Ocean Council:* Executive Order 13547 regarding Stewardship of the Ocean, Our Coasts, and the Great Lakes established a national policy to ensure the protection, maintenance, and restoration of the health of ocean, coastal, and Great Lakes ecosystems, among other things. The order also created a National Ocean Council to ensure that federal agency decisions and actions affecting the ocean, coasts, and Great Lakes will be guided by articulated stewardship principles and national priority objectives.[90]

- *Reducing Emissions from Deforestation and Forest Degradation (REDD+):* In December 2010, the U.S. government released its strategy to reduce emissions from deforestation and forest degradation and increase carbon sequestration by forests in developing countries. This governmentwide strategy outlines how the United States will allocate and invest the $1 billion dedicated for REDD+ announced at the 2009 meeting of the parties to the United Nations Framework Convention on Climate Change in Copenhagen.[91]
- *Recovery through Retrofit:* According to CEQ, they facilitated a broad interagency process to propose federal action that would expand green job opportunities in the United States and boost energy savings by improving home energy efficiency. The resulting report released in October 2009 builds on investments made in the Recovery Act to expand the home energy efficiency and retrofit market.[92]

Regulations and Guidance

While Congress has not passed comprehensive legislation intended to reduce greenhouse gas emissions, there are several ongoing federal efforts to develop regulations and guidance related to climate change. These include, but are not limited to, (1) Environmental Protection Agency (EPA) regulations under the Clean Air Act, (2) EPA greenhouse gas reporting rules, and (3) draft CEQ guidance to agencies on the consideration of greenhouse gases when fulfilling the requirements of the National Environmental Policy Act of 1969 (NEPA).

In 2007, the Supreme Court ruled that EPA has the statutory authority to regulate greenhouse gas emissions from new motor vehicles under the Clean Air Act because greenhouse gases meet the act's definition of an air pollutant. Furthermore, the Supreme Court held that EPA must regulate greenhouse gas emissions as an air pollutant if EPA finds them to be an endangerment to public health or welfare. In response to this case, EPA issued a finding that carbon dioxide, methane, nitrous oxide, and hydrofluorocarbon emissions from new motor vehicles are contributing to air pollution, which is endangering public health and welfare.[93] Based on this endangerment finding,

EPA issued a final rule establishing greenhouse gas emissions standards for new light-duty motor vehicles on May 7, 2010.[94] Under EPA's current interpretation of the Clean Air Act, greenhouse gas emissions from certain stationary sources will be subject to regulation under the act's Prevention of Significant Deterioration provisions beginning in 2011 as a result of this final rule.[95]

However, in May 2010, EPA issued a final rule that would only impose the Clean Air Act's Prevention of Significant Deterioration and Title V permitting provisions on a select number of stationary sources, including coal-fired power plants, beginning in 2011. Under this rule, if a coal-fired power plant is built or an existing plant makes a major modification—a physical or operational change that would result in a significant net increase in emissions—the plant would need to obtain a Prevention of Significant Deterioration permit from the appropriate regulatory authority that implements the best available control technology for greenhouse gas emissions.[96]

The Consolidated Appropriations Act of 2008 directed EPA to issue a regulation requiring mandatory reporting of greenhouse gas emissions above appropriate thresholds in all sectors of the economy. In accordance with the direction in the joint explanatory statement accompanying the act, EPA issued the regulation under its Clean Air Act authority on October 30, 2009. The regulation requires certain facilities that directly emit greenhouse gases and upstream suppliers of fossil fuels and industrial greenhouse gases, as well as manufacturers of vehicles and engines, to report their annual greenhouse gas emissions. Specifically, facilities that emit 25,000 metric tons or more of carbon dioxide equivalent greenhouse gas emissions per year and most upstream suppliers and vehicle and engine manufacturers are required to report their emissions. The regulation includes provisions to ensure the accuracy of emissions data through monitoring, recordkeeping, and verification requirements.

On February 18, 2010, CEQ proposed changes to how federal agencies implement the NEPA, in conjunction with its fortieth anniversary.[97] CEQ released for public comment draft guidance on how federal agencies should analyze the environmental effects of greenhouse gas emissions and climate change when they describe the environmental effects of a proposed action under NEPA.[98] The draft guidance advises agencies that a quantitative and qualitative assessment of greenhouse gas emissions from a proposed action that would be reasonably anticipated to cause direct emissions of 25,000 metric tons of carbon dioxide equivalent or more may provide meaningful information to decision makers and the public. Once the agency determines that an assessment of greenhouse gas emissions or climate change issues is appropriate, the draft guidance further instructs the agency to identify alternative actions that are both adapted to anticipated climate change impacts and mitigate the emissions that cause climate change.

International Commitments

The December 2009 Copenhagen Accord, a nonbinding political agreement, articulated a collective commitment by developed countries to provide new and additional resources approaching $30 billion between 2010 and 2012 to support developing countries' mitigation and adaptation efforts, according to a November 2010 statement by the Department of State. In accordance with the accord, the current Administration proposed that the United States would reduce greenhouse gas emissions in the range of 17 percent below 2005 levels by 2020 in conformity with anticipated energy and climate legislation.[99] According to an April 2010 Department of State fact sheet, federal agencies are working with international partners to provide "fast start" climate finance approaching $30 billion during the period of 2010 to 2012 to help meet the adaptation and mitigation needs of developing countries. Programs identified as priorities include reducing emissions from deforestation and degradation and efforts to deploy clean energy technologies in developing countries.[100]

Congressional Testimony

Two 2009 testimonies by Dr. John P. Holdren, Director of OSTP, provide a governmentwide view of climate change programs and priorities in the current Administration. In a July 30, 2009, testimony, Dr. Holdren emphasized three areas needing comprehensive and coordinated treatment from USGCRP: (1) adaptation research to increase knowledge about the abilities of communities, regions, and sectors to adapt to a changing climate; (2) integrated assessment of how regional and local climate impacts are experienced in different ways in different places across the country and across different economic sectors; and (3) climate services and coordinated information to assist decision making across public and private sectors, such as providing local planners with information on likely changes in precipitation amounts and flooding rains.[101] Similarly, in a December 2, 2009, testimony,

Dr. Holdren stated, "Besides enhancing research and modeling of the physical climate system, four areas of particular need for more comprehensive and coordinated treatment from USGCRP are Earth observations, adaptation research, integrated assessment, and climate services."[102]

APPENDIX VIII. SUPPLEMENT ON HOW FEDERAL CLIMATE ACTIVITIES ARE COORDINATED, PROVIDED BY CEQ, OMB, AND OSTP IN TECHNICAL COMMENTS

In technical comments, CEQ, OMB, and OSTP submitted table 3 to display how climate change is a complex, crosscutting issue, where many federal entities manage related program and activities. According to CEQ, OMB, and OSTP, this supplement provides a high-level representation of how informal processes and ad hoc meetings play a key role in setting climate priorities in addition to the formal processes described in figure 4 in the body of this report.

Table 3. Selected Examples of Mechanisms for Coordination on Federal Climate Change Activities Provided by CEQ, OMB, and OSTP in Technical Comments

Category	Selected examples
Cabinet-level processes	Green/energy cabinet meetings National Ocean Council
National Science and Technology Council	Committee on Environment, Natural Resources, and Sustainability: U.S. Global Change Research Program Roundtable on Climate Information and Services Task Force on Integrating Science and Technology for Sustainability Subcommittee on Water Availability and Quality Arctic Research Policy Committee
Issue-specific task forces, programs, and initiatives	Climate Change Adaptation Task Force, Interagency Task Force on Carbon Capture and Storage, Climate Change Technology Program, Executive Order on Federal Leadership in Environmental, Energy, and Economic Performance
Interagency processes	Legislative Referral Memorandum (LRM) process , Review of proposed rules under Executive Order 12866, President's Annual budget process

Source: CEQ, OMB, and OSTP technical comments

APPENDIX IX. GAO CLIMATE CHANGE FUNDING WEB-BASED QUESTIONNAIRE

This appendix presents the text of the Web-based questionnaire that we developed to gather information and opinions of key federal officials involved in defining and reporting climate change funding, developing strategic climate change priorities, or aligning funding with strategic priorities. Several hyperlinks to additional information embedded within the questionnaire are not reproduced in this appendix.

Introduction

The U.S. Government Accountability Office (GAO), an independent, nonpartisan congressional agency, is studying federal funding for climate change programs and activities at the request of Representative Edward Markey, Chairman of both the House Select Committee on Energy Independence and Global Warming and the Subcommittee on Energy and Environment, Committee on Energy and Commerce. Through this web- based questionnaire, GAO is seeking insights from key federal officials, such as you, to the following questions:

1. To what extent do federal agencies use a consistent methodology for defining and reporting climate change funding?
2. What are federal strategic climate change priorities, and to what extent is funding aligned with these priorities?
3. What options, if any, are available to better align federal climate change funding with strategic priorities?

Your responses to the questionnaire will help GAO explain to Congress how federal climate change funding is tracked and directed toward goals. This information may assist Congress as it considers climate change policy options.

This questionnaire is divided into five sections: (I) Background, (II) Defining and Reporting Federal Climate Change Funding, (III) Setting Federal Strategic Climate Change Priorities, (IV) Aligning Federal Climate Change Funding with Strategic Priorities, and (V) Conclusion. Sections II, III, and IV begin with questions asking about your familiarity with the section topics. Because your experience and knowledge about these topics differ from that of other respondents, the questionnaire allows you to only answer questions that are related to topics with which you are most familiar.

Timing

The questionnaire can be filled out in as little as 30 minutes, but may take longer depending on the amount of information you would like to share in response to open-ended questions. We understand that there are great demands on your time, but your response is critical to help us provide information to Congress.

Your Responses

Please respond to the questions to the best of your knowledge and experience without consulting others. Our report to Congress will not associate your name with your responses or list you as a respondent. Your responses will be combined with those of the other respondents and summarized in our report to Congress.

Navigation, Exiting and Printing the Survey

To save and exit: you can complete the questionnaire in one or multiple sessions. To end a session, or upon completion of the questionnaire, click on the "Save and Exit" button at the bottom of the screen. Always use the "Save and exit" button to close the questionnaire. If you do not use this button, You will lose all the responses entered on the last screen that you viewed. To paste your responses from another document: select the text you would like to paste, copy it, click on the question text box, and type "Ctrl" and "V" to paste the text in the box. To learn more about navigating, exiting, and printing the survey, please click here.

To get help: Should you have any questions, please click here.

Abbreviations

CCTP - Climate Change Technology Program
CEQ - Council on Environmental Quality

GAO - Government Accountability Office
OMB - Office of Management and Budget
OSTP - Office of Science and Technology Policy
USGCRP - United States Global Change Research Program

Key Terms

Approaches for aligning climate change funding with priorities means systems utilized to clearly define how funding will be used to achieve goals. These could include formal strategic plans or memos or other guidance issued through the budget process.

Climate Change Funding - Click here for some basic information on how the Office of Management and Budget presents climate change funding in its reports to Congress.

Definitions may include OMB Circular A-11, which describes the types of programs that agencies should report to OMB as part of the Climate Change Technology Program (CCTP), United States Global Change Research Program (USGCRP), and climate change international assistance. Definitions may also include more specific guidance such as the Climate Change Technology Program Classification Criteria found in CCTP's 2006 Strategic Plan.

Method means practices or procedures used to define and report climate change funding.

Organization means federal agencies, sub-agencies, or departments, Executive Office of the President entities such as CEQ, and crosscutting programs such as USGCRP.

Reporting refers to how funding information makes its way from the program level through federal agencies to OMB. Reporting can take place at several stages in the budget process and may involve systems such as OMB's MAX.

I. Background

1. Please provide your name, title, and organization.
2. Which best describes the location of your position within the federal government?
   - Executive Office of the President, such as OMB, OSTP, and CEQ
   - Crosscutting program, such as USGCRP and CCTP
   - Agency, subagency, or department
   - Multiple (please explain)
   - Other (please explain)

   If you chose 'multiple', please explain:
   If you chose 'other', please explain:
3. Which best describes your position?
   - Primarily budget oriented
   - Primarily policy oriented
   - Both budget and policy oriented
   - Program manager
   - Scientist
   - Other (please explain)

   If you chose 'other', please explain:
4. Within the last fiscal year, approximately what percentage of your work time was spent on tasks directly or indirectly related to climate change?

II. Defining and Reporting Federal Climate Change Funding

In the following questions, we ask about how funding for climate change programs and activities is defined and reported within your organization and across the federal government.

Your Organization

5. Do you have sufficient knowledge or experience to answer the following questions about the methods employed by your organization to define and report climate change funding? (Before responding, please scroll down to review questions 6 to 11)

- Yes (Go to question 6)
- No (Click here to skip to question 12.)
- Not applicable (Click here to skip to question 12.)

6. How does your organization define and report climate change funding? (Please be specific about any definitions, guidance, or any other methods you refer to in your response)
7. To what extent does your organization consistently define and report climate change funding from year to year?
8. To what extent are current methods for defining and reporting climate change funding applied consistently within your organization?
9. What are the strengths of current methods employed by your organization to define and report climate change funding, if any?
10. What are the limitations of current methods employed by your organization to define and report climate change funding, if any?
11. If you identified limitations with current methods in question 10, how might they be addressed?

Across the Federal Government

12. Do you have sufficient knowledge or experience to answer the following questions about the methods employed to define and report climate change funding across the federal government? (Before responding, please scroll down to review questions 13 to 17).
    - Yes (Go to question 13)
    - No (Click here to skip to question 18.)
13. To what extent is climate change funding defined and reported consistently across the federal government from year to year?
14. To what extent do officials across the federal government consistently apply current methods for defining and reporting climate change funding? Please be specific about which methods you are referring to in your responses.
15. What are the strengths of current methods for defining and reporting climate change funding across the federal government, if any? Please be specific about which methods you are referring to in your responses.
16. What are the limitations of current methods for defining and reporting climate change funding across the federal government, if any? Please be specific about which methods you are referring to in your responses.

17. If you identified limitations with current methods in question 16, how might they be addressed?

III. Setting Federal Strategic Climate Change Priorities

In the following questions, we ask about how strategic priorities for climate change programs and activities are set across the federal government.

18. Do you have sufficient knowledge or experience to answer the following questions about how the federal government sets strategic climate change priorities? (Before responding, please scroll down to review questions 19 – 24).
    - Yes (Go to question 19)
    - No (Click here to skip to question 25.)
19. What are the current strategic climate change priorities across the federal government?
20. What are the current processes for setting strategic climate change priorities across the federal government?
21. What are the strengths of the current processes for setting strategic climate change priorities across the federal government, if any?
22. What are the limitations of the current processes for setting strategic climate change priorities across the federal government, if any?
23. If you identified limitations with the current processes for setting priorities in question 22, how might they be addressed?
24. If you identified limitations with the current processes for setting priorities in question 22, who within the federal government should take the lead on addressing them? Please explain your rationale.

IV. Aligning Funding with Federal Strategic Climate Change Priorities

In the following questions, we ask about how funding for climate change programs and activities is aligned with federal strategic priorities within your organization and across the federal government.

Your Organization

25. Do you have sufficient knowledge or experience to answer the following questions about how your organization aligns its climate change funding with federal strategic priorities? (Before responding, please scroll down to review questions 26 - 29)
    - Yes (Go to question 26)
    - No (Click here to skip to question 30.)
    - Not applicable (Click here to skip to question 30.).
26. To what extent does your organization's planning and funding reflect strategic priorities across the federal government?
27. What are the strengths of your organization's current approaches for aligning its climate change funding with federal strategic priorities?
28. What are the limitations of your organization's current approaches for aligning its climate change funding with federal strategic priorities?
29. If you identified limitations with your organization's approaches in question 28, how might they be addressed?

Across the Federal Government

30. Do you have sufficient knowledge or experience to answer the following questions about how climate change funding is aligned with strategic priorities across the federal government? (Before responding, please scroll down to review questions 31 - 35)
    - Yes (go to question 31)
    - No (Click here to skip to question 36.)
31. To what extent is climate change funding aligned with strategic priorities across the federal government?
32. What are the strengths of current approaches for aligning climate change funding with strategic priorities across the federal government, if any?
33. What are the limitations of current approaches for aligning climate change funding with strategic priorities across the federal government, if any?
34. If you identified limitations in question 33, how might they be addressed?
35. If you identified limitations in question 33, who within the federal government should take the lead on addressing them? Please explain your rationale.

CONCLUSION

Please share any additional comments.

End Notes

[1] USGCRP coordinates and integrates federal research on changes in the global environment and their implications for society.

[2] *Global Climate Change Impacts in the United States*, Thomas R. Karl, Jerry M. Melillo, and Thomas C. Peterson, eds. (Cambridge University Press, 2009).

[3] Office of Management and Budget, *Federal Climate `Change Expenditures Report to Congress* (June 2010). See http://www.whitehouse.gov/sites/default/files/omb/assets/ legislative_ reports/FY2011_Climate_Change.pdf

[4] Congressional Research Service, *Climate Change: Federal Program Funding and Tax Incentive*, RL33817 (Dec. 22, 2008).

[5] GAO, *Environmental Satellites: Strategy Needed to Sustain Critical Climate and Space Weather Measurements*, GAO-10-456 (Washington, D.C.: Apr. 27, 2010). GAO, *Climate Change Adaptation: Strategic Federal Planning Could Help Federal Officials Make More Informed Decisions*, GAO-10-113 (Washington, D.C.: Oct. 7, 2009). Climate change adaptation means adjustments to natural or human systems in response to actual or expected climate change.

[6] GAO, *Results-Oriented Government: Practices That Can Help Enhance and Sustain Collaboration among Federal Agencies*, GAO-06-15 (Washington, D.C.: Oct. 21, 2005), and *Managing for Results: Barriers to Interagency Coordination*, GAO/GGD-00-106 (Washington, D.C.: Mar. 29, 2000).

[7] We sent our questionnaire to several OSTP officials, asking that they respond individually, but OSTP elected to provide a single formal response. Throughout this report we attribute information from this formal collective response directly to OSTP using phrases such as "according to OSTP" or "OSTP stated." We count the collective OSTP response as one of the 73 included in the 69 percent response rate.

[8] GAO, *Climate Change: Federal Reports on Climate Change Funding Should Be Clearer and More Complete*, GAO-05-461 (Washington, D.C.: Aug. 25, 2005).

[9] The revenue losses resulting from provisions of federal tax laws may, in effect, be viewed as expenditures channeled through the tax system. Like the annual lists of tax expenditures prepared by the Department of the Treasury, this report considers only tax expenditures related to individual and corporate income taxes and does not address excise taxes.

[10] OMB's *Fiscal Year 2007 Report to Congress on Federal Climate Change Expenditures* (April 2006) is available at http://www.whitehouse.gov/sites/default/files/ omb/assets/omb/ legislative/fy07_climate_change.pdf. See appendix II for information about these OMB reports.

[11] See appendixes III through VI for detailed climate change funding data from OMB reports.

[12] Pub. L. No. 111-5 (2009). Certain energy grants in lieu of new technology or energy investment tax credits authorized by the Recovery Act are accounted for in the tax expenditures category.

[13] These categories are reported separately following this section of the report and in appendixes V and VI.

[14] Each year from 1998 to 2007, an annual appropriations law required OMB to report certain climate change funding data. Annual appropriations laws for 2008 and 2009 did not require OMB to report such data, but the annual appropriations law for 2010 did require OMB to

report on certain climate change funding for 2009 and 2010. OMB's most recent report was released in June 2010. For more information about recent OMB climate change funding reports, see appendix II.

[15] GAO-05-461.

[16] Pub. L. No. 111-5 (2009). Recovery Act funding was available for obligation until September 30, 2010. This sum does not include $3.1 billion for energy grants in lieu of new technology or energy investment tax credits available under the Recovery Act. This $3.1 billion is accounted for in the tax expenditures category discussed in this section.

[17] In technical comments, CEQ, OMB, and OSTP noted that Recovery Act funding was primarily intended to stimulate the economy, not necessarily address climate change. The EOP entities noted that co-benefits are a big part of technology funding, citing weatherization funding in the Recovery Act as an example in which there is a climate change benefit from efficiency, but which also helps low-income Americans disproportionately impacted by higher energy prices.

[18] GAO, *Department of Energy: Further Actions Are Needed to Improve DOE's Ability to Evaluate and Implement the Loan Guarantee Program*, GAO-10-627 (Washington, D.C.: July 12, 2010). Other relevant GAO reports include *Independent Oversight of Recovery Act Funding for Mississippi's Weatherization Assistance Program*, GAO-10-796R (Washington, D.C.: June 30, 2010) and *Recovery Act: Opportunities to Improve Management and Strengthen Accountability over States' and Localities' Uses of Funds*, GAO-10-999 (Washington, D.C.: Sept. 20, 2010).

[19] The revenue losses resulting from provisions of federal tax laws may, in effect, be viewed as expenditures channeled through the tax system. Like the annual lists of tax expenditures prepared by the Department of the Treasury, this report considers only tax expenditures related to individual and corporate income taxes and does not address excise taxes.

[20] OMB did not report revenue loss estimates for existing climate-related tax expenditures from 1993 through 2004.

[21] Panel of the National Academy of Public Administration, *Building Strong for Tomorrow: NOAA Climate Service*, a report prepared for Congress, the Department of Commerce, and NOAA (Sept. 13, 2010).

[22] In March 2011, the Office of Energy and Climate Change Policy joined the Domestic Policy Council.

[23] The Departments of Agriculture, Commerce, Defense, Energy, the Interior, State, Transportation, and Health and Human Services participate in both USGCRP and CCTP. The Environmental Protection Agency, National Aeronautics and Space Administration, National Science Foundation, and the U.S. Agency for International Development also participate in both USGCRP and CCTP. The Smithsonian Institution participates in USGCRP but not in CCTP.

[24] Pub. L. No. 101-606 (1990). For more information about USGCRP, see http://www.globalchange.gov/.

[25] Pub. L. No. 109-58, § 1601 (2005).

[26] For detailed information about the federal budget process, see GAO, *A Glossary of Terms Used in the Federal Budget Process*, GAO-05-734SP (Washington, D.C.: September 2005).

[27] U.S. Global Change Research Program and the Subcommittee on Global Change Research, *Our Changing Planet* (Washington, D.C., October 2009).

[28] In our Web-based questionnaire, we defined "method" as practices or procedures used to define and report climate change funding, and "reporting" as how funding information makes its way from the program level through federal agencies to OMB.

[29] For further information about the climate change components of OMB Circular A-11, see Section 84—Character Classification (Schedule C) at http://www.whitehouse.gov/omb/circulars_a11_current_year_a11_toc/. In its reports to the Congress, OMB also presents tax expenditures related to climate change. OMB works with the Department of the Treasury to identify relevant tax expenditures.

[30] See https://max.omb.gov/maxportal/ for more information about OMB's MAX system.

[31] CEQ, OMB, and OSTP submitted consolidated technical comments on this report.

[32] For more information about CCTP's only Strategic Plan to date, see http://www.climate technology.gov/.

[33] GAO, *Climate Change: A Coordinated Strategy Could Focus Federal Geoengineering Research and Inform Governance Efforts*, GAO-10-903 (Washington, D.C.: Sept. 23, 2010).

[34] As previously noted, we attribute information from OSTP's formal collective response to our questionnaire directly to OSTP using phrases such as "according to OSTP" or "OSTP stated."

[35] USGCRP budget data are documented and tabulated in its *Our Changing Planet* annual report to Congress.

[36] Congressional Research Service, *Climate Change: Federal Program Funding and Tax Incentives*, RL33817 (Dec. 22, 2008).

[37] OMB stopped reporting this distinction in 2001. See appendix V for more details.

[38] Congressional Budget Office, *Federal Climate Change Programs: Funding History and Policy Issues*, Pub. No. 4025 (Washington, D.C., March 2010)

[39] Congressional Research Service, *Climate Change: Federal Program Funding and Tax Incentives*, RL33817 (Dec. 22, 2008).

[40] U.S. Congress, Office of Technology Assessment, *Preparing for an Uncertain Climate—Volume I*, OTA-O-567 (Washington, D.C.: U.S. Government Printing Office, October 1993).

[41] CEQ, OMB, and OSTP technical comments emphasized the importance of addressing co-benefits in greater detail, noting that, for example, research and development support for clean energy technologies has clear economic benefits as it addresses a market failure for underinvestment in early stage research and development. Additionally, they noted investments in alternative fuel vehicles have a security impact because they support reducing America's oil consumption.

[42] Office of Management and Budget and Office of Science and Technology Policy, *Science and Technology Priorities for the FY 2012 Budget*, Memorandum for the Heads of Executive Departments and Agencies, M-10-30 (July 21, 2010), available at http://www.white-house.gov/administration/eop/ostp/rdbudgets.

[43] This supplemental guidance is available at http://www.whitehouse.gov/administration/eop/ ostp/rdbudgets.

[44] Office of Management and Budget and Office of Science Technology Policy, *Science and Technology Priorities for the FY 2011 Budget*, Memorandum for the Heads of Executive Departments and Agencies, M-09-27 (Aug. 4, 2009), available at http://www.white house.gov/omb/asset.aspx?AssetId=1565.

[45] Congressional Research Service, *Climate Change: Federal Program Funding and Tax Incentives*, RL33817 (Dec. 22, 2008).

[46] USGCRP is undertaking a strategic planning process to ensure the alignment of administration priorities, National Research Council guidance, and agency programs and budgets, according to CEQ, OMB, and OSTP.

[47] GAO, *Global Warming: Administration Approach Cautious Pending Validation of Threat*, GAO/NSIAD-90-63 (Washington, D.C.: Jan. 8, 1990).

[48] GAO-06-15 and *Managing for Results: Barriers to Interagency Coordination*, GAO/GGD-00-106 (Washington, D.C.: Mar. 29, 2000).

[49] When provided with this response, CEQ, OMB, and OSTP clarified that this statement referred to the USGCRP coordination office, not the program.

[50] National Academies, National Research Council, *Restructuring Federal Climate Research toMeet the Challenges of Climate Change* (Washington D.C.: The National Academies Press, 2009).

[51] See http://www.climatescience.gov/Library/stratoptions/climate-professional-Oct08.pdf to access the USGCRP Climate Experts Roundtable listening session.

[52] Since 1994, we have recommended integrated budget reviews of tax expenditures and related spending programs to determine the most effective methods for achieving federal objectives. See GAO, *Tax Policy: Tax Expenditures Deserve More Scrutiny*, GAO/GGD/AIMD-94-122 (Washington, D.C.: June 3, 1994) and *Government Performance and Accountability: Tax Expenditures Represent a Substantial Federal Commitment and Need to Be Reexamined*, GAO-05-690 (Washington, D.C.: Sept. 23, 2005).

[53] GAO-06-15 and GAO/GGD-00-106.

[54] In addition, such a governmentwide strategic plan would seem consistent with the recently enacted GPRA Modernization Act of 2010 (Pub. L. No. 111-352 (2011)), which requires OMB to coordinate with agencies to develop priority goals to improve the performance and management of the federal government, including outcome-oriented goals covering a limited number of crosscutting policy areas. In addition, the act requires OMB, in coordination with other federal agencies, to develop the federal government performance plan. This plan must establish performance goals for the federal government, including performance goals related to the priority goals, identify major management challenges that are governmentwide or crosscutting in nature, and describe plans to address such challenges, among other things. For each federal government performance goal, the plan must identify the agencies, organizations, program activities, regulations, tax expenditures, policies, and other activities contributing to the performance goal.

[55] Panel of the National Academy of Public Administration, *Building Strong for Tomorrow: NOAA Climate Service*, a report prepared for Congress, the Department of Commerce, and NOAA (Sept. 13, 2010).

[56] Panel of the National Academy of Public Administration, *Building Strong for Tomorrow: NOAA Climate Service*, a report prepared for Congress, the Department of Commerce, and NOAA (Sept. 13, 2010).

[57] See http://www.climatescience.gov/Library/stratoptions/climate-professional-Oct08.pdf to access the USGCRP Climate Experts Roundtable listening session.

[58] See Roger A. Pielke, Jr., *The Development of the U.S. Global Change Research Program, 1987-1994*, a policy case study prepared for the 2001 American Meteorological Society Policy Symposium (Environmental and Societal Impacts Group, National Center for Atmospheric Research, May 2001). See also W. Henry Lambright, "The Rise and Fall of Interagency Cooperation: The U. S. Global Change Research Program," *Public Administration Review*, vol. 57, no. 1 (Jan.-Feb. 1997), 36-44.

[59] National Academies, National Research Council, *Restructuring Federal Climate Research toMeet the Challenges of Climate Change* (Washington D.C.: The National Academies Press, 2009).

[60] Statement of Dr. John H. Marburger, III, Director, Office of Science and Technology Policy to the Committee on Commerce, Science and Transportation, United States Senate, *A Time for Change: Improving the Federal Climate Change Research and Information Program* (Nov, 14, 2007).

[61] Office of Management and Budget and Office of Science and Technology Policy, *Science and Technology Priorities for the FY 2012 Budget*, Memorandum for the Heads of Executive Departments and Agencies, M-10-30 (July 21, 2010), available at http://www.whitehouse.gov/administration/eop/ostp/rdbudgets.

[62] Panel of the National Academy of Public Administration, *Building Strong for Tomorrow: NOAA Climate Service*, a report prepared for Congress, the Department of Commerce, and NOAA (Sept. 13, 2010).

[63] According to technical comments from CEQ, OMB, and OSTP, the purpose of the Roundtable on Climate Information and Services is to bring together those CENRS members who have a mandate, obligation, and/or interest in the development and provision of climate services,

including supporting science and technology, to rationalize the functional, operational, and structural elements of climate services for the nation.

[64] For more information on USGCRP's annual reports to Congress, see http://www.globalchange.gov/publications/our-changing-planet-ocp.

[65] Congressional Budget Office, *Federal Climate Change Programs: Funding History and Policy Issues*, Pub. No. 4025 (Washington, D.C., March 2010), and Congressional Research Service, *Climate Change: Federal Program Funding and Tax Incentives*, RL33817 (Dec. 22, 2008).

[66] CEQ, OMB, and OSTP technical comments noted that the examples cited are not all of the mechanisms used to set federal climate change policies and that there is a difference between policy mechanisms and science research efforts.

[67] *Strategic Plan for the U.S. Climate Change Science Program* (July 2003), http://www.globalchange.gov/about/strategic-plan-2003/2003-strategic-plan.

[68] OMB's *Fiscal Year 2011 Report to Congress on Federal Climate Change Expenditures* (June 2010) is available at http://www.whitehouse.gov/omb/asset.aspx?AssetId=2776.

[69] Revised research plan for the U.S. Climate Change Science Program (May 2008), http://www.globalchange.gov/about/strategic-plan-2003/revised-research-plan.

[70] See http://www.globalchange.gov/publications/reports/nrc-reports for key reports from the National Research Council (NRC) that are relevant to USGCRP's planning and evolution. The NRC functions under the auspices of the National Academy of Sciences, the National Academy of Engineering, and the Institute of Medicine. The mission of the NRC is to improve government decision making and public policy, increase public education and understanding, and promote the acquisition and dissemination of knowledge in matters involving science, engineering, technology, and health.

[71] The Consolidated Appropriations Act of 2008 (Pub. L. No. 110-161(2007)) required the National Oceanic and Atmospheric Administration to enter into an agreement with the National Academy of Sciences under which the latter would establish a committee to "investigate and study the serious and sweeping issues relating to global climate change and make recommendations regarding what steps must be taken and what strategies must be adopted in response to global climate change, including the science and technology challenges thereof." To fulfill this mandate, the National Academies completed a series of reports collectively titled *America's Climate Choices* available at http://americasclimatechoices.org/.

[72] *U.S. Climate Change Technology Program Strategic Plan*, DOE/PI-0005 (September 2006), http://www.climatetechnology.gov/stratplan/final/CCTP-StratPlan-Sep-2006.pdf.

[73] Geoengineering means large-scale deliberate interventions in the earth's climate system to diminish climate change or its impacts.

[74] United States Department of Agriculture, *Strategic Plan for FY 2010-2015*, http://www.ocfo.usda.gov/usdasp/sp2010/sp2010.pdf.

[75] Office of Management and Budget and Office of Science and Technology Policy, *Science and Technology Priorities for the FY 2012 Budget*, Memorandum for the Heads of Executive Departments and Agencies, M-10-30 (July 21, 2010), available at http://www.whitehouse.gov/administration/eop/ostp/rdbudgets.

[76] This supplemental guidance is available at http://www.whitehouse.gov/administration/eop/ostp/rdbudgets.

[77] Office of Management and Budget and Office of Science and Technology Policy, Memorandum for the Heads of Executive Departments and Agencies, *Science and Technology Priorities for the FY 2011 Budget*, M-09-27 (Aug. 4, 2009), available at http://www.whitehouse.gov/omb/asset.aspx?AssetId=1565.

[78] Additional information on the October 5, 2009, Executive Order 13514 on Federal Leadership in Environmental, Energy, and Economic Performance is available at http://www.whitehouse.gov/administration/eop/ceq/sustainability.

[79] Federal Agency Strategic Sustainability Performance Plans are available at http://www.whitehouse.gov/administration/eop/ceq/sustainability/plans.
[80] For more information about the Climate Change Adaptation Task Force, see http://www.whitehouse.gov/administration/eop/ceq/initiatives/adaptation.
[81] The White House Council on Environmental Quality, *Progress Report of the Interagency Climate Change Adaptation Task Force: Recommended Actions in Support of a National Climate Change Adaptation Strategy* (October 5, 2010). This report is available at http://www.whitehouse.gov/sites/default/files/microsites/ceq/Interagency-Climate-Change-A daptation-Progress-Report.pdf.
[82] OMB's Fiscal Year 2011 Report to Congress on Federal Climate Change Expenditures (June 2010) is available at http://www.whitehouse.gov/omb/asset.aspx?AssetId=2776.
[83] For more information about the Interagency Task Force on Carbon Capture and Storage, see http://www.whitehouse.gov/administration/eop/ceq/initiatives/ccs.
[84] The Carbon Capture and Storage Task Force report is available on the Department of Energy's Web site at http://www.fe.doe.gov/programs/sequestration/ccs_task_force.html and the Environmental Protection Agency's Web site at http://www.epa.gov/climatechange/policy/ccs_task_force.html.
[85] Pub. L. No. 10 1-606, § 106 (1990), *codified at* 15 U.S.C. § 2936.
[86] For more information about the National Climate Assessment, including links to newsletters, task force meetings, and workshops, see http://www.globalchange.gov/what-we-do/assessment.
[87] 75 Fed. Reg. 54403 (Sept. 7, 2010).
[88] For more information about the NOAA Climate Service, see http://www.noaa.gov/climate.html. A range of climate information is presented at www.climate.gov, NOAA's Climate Services Portal.
[89] The Department of Defense and Full Year Continuing Appropriations Act, 2011 prohibits any funds appropriated in the act to be used to implement, establish, or create a NOAA Climate Service as NOAA had previously described it.
[90] For more information about the National Ocean Council, see http://www.whitehouse.gov/administration/eop/oceans.
[91] For more information about the REDD+ Strategy, see http://www.usaid.gov/our_work/environment/climate/policies_prog/redd.html.
[92] For more information about the Recovery Through Retrofit report, see http://www.whitehouse.gov/administration/eop/ceq/initiatives/retrofit.

[93] A bill has been introduced which would exclude greenhouse gases from being defined as "air pollutants" subject to regulation under the Clean Air Act. *See* H.R. 97, 112th Cong. (2011). In addition, numerous lawsuits have been filed challenging the endangerment finding, which have been consolidated into one case in the D.C. Circuit Court of Appeals.
[94] Numerous lawsuits challenging this rule have been filed and consolidated into one case in the D.C. Circuit Court of Appeals.
[95] *See* 75 Fed. Reg. 17,004 (Apr. 2, 2001). Known as the timing or triggering rule, numerous lawsuits have been filed challenging this rule.
[96] 75 Fed. Reg. 31514 (June 3, 2010). Known as the tailoring rule, numerous lawsuits have been filed challenging this rule. These lawsuits have been consolidated with the lawsuits challenging the timing or triggering rule in the D.C. Circuit Court of Appeals. In addition, bills have been introduced which would preclude EPA from regulating the greenhouse gas emissions of stationary sources under the Clean Air Act. *See* H.R. 153 (112th Cong.); H.R. 199 (112th Cong.); S. 228 (112th Cong.); S. 231 (112th Cong.).
[97] Pub. L. No. 91-190 (1970). For more information on CEQ's draft NEPA guidance, see http://www.whitehouse.gov/administration/eop/ceq/initiatives/nepa.
[98] The guidance is not applicable to federal land and resource management actions.
[99] See http://unfccc.int/home/items/5264.php for more information.

[100] For more information about international climate change finance, see the Department of State Climate Change Web page at http://www.state.gov/g/oes/climate/.

[101] Statement of Dr. John P. Holdren, Director, Office of Science and Technology Policy, Executive Office of the President, before The Committee on Commerce, Science and Transportation, United States Senate (July 30, 2009), http://www.whitehouse.gov/ galleries/press_release_files/HoldrenTestimony.pdf.

[102] Testimony of John P. Holdren, Assistant to the President for Science and Technology and Director of the Office of Science and Technology Policy, Executive Office of the President, before the Select Committee on Energy Independence and Global Warming, U.S. House of Representatives, on the Administration's View of the State of the Climate (December 2, 2009), http://www.whitehouse.gov/administration/eop/ ostp/pressroom/12022009.P

In: Climate Change
Eds: A.H. Anderson and D. J. Thomas
ISBN: 978-1-61942-760-0

Chapter 2

FISCAL YEAR 2011 REPORT TO CONGRESS ON FEDERAL CLIMATE CHANGE EXPENDITURES*

Report to Congress

1. INTRODUCTION

> "So we have a choice to make. We can remain one of the world's leading importers of foreign oil, or we can make the investments that would allow us to become the world's leading exporter of renewable energy. We can let climate change continue to go unchecked, or we can help stop it. We can let the jobs of tomorrow be created abroad, or we can create those jobs right here in America and lay the foundation for lasting prosperity."
>
> -President Barack Obama, March 19, 2009.

The following is an accounting of Federal funding for climate change programs and activities, both domestic and international, included in the President's fiscal year 2011 Budget. This report is provided in response to Title IV, Section 426 of Public Law 111-88, the Department of the Interior, Environment, and Related Agencies Appropriations Act of 2010.

---

* This is an edited, reformatted and augmented version of a Federal Climate Change Expenditures Report to Congress publication, dated June, 2010.

1.1. Background

The U.S. Government's portfolio of climate change programs and cross-cutting initiatives focuses on reducing the fundamental scientific uncertainties associated with climate change; advancing the development and introduction of energy-efficient, renewable, and other low- or non-emitting technologies; improving standards for measuring and registering emissions reductions and supporting adaptation and resilience to climate change impacts. Many elements of the Administration's climate change portfolio are designed to provide incentives for greenhouse gas emissions reductions throughout the United States and help developing nations or regions to do the same or adapt to the effects of climate change.

In February 2009, President Obama announced his intent to work with Congress in seeking new legislation on energy and climate change that would establish a mandatory economy-wide cap on emissions, with emission reductions beginning in 2012 and becoming more stringent annually thereafter, leading to greenhouse gas (GHG) reductions of approximately 83 percent below 2005 levels by 2050. In June 2009, the U.S. House of Representatives took a significant step toward realizing this goal by passing the American Clean Energy and Security Act. The legislation includes major new investments in clean energy technologies that will be needed to achieve a prosperous transformation to a low-carbon economy, help the United States adapt to the effects of climate change, and establish the United States as a leader in creating the skilled green jobs that will help drive economic growth in coming decades.

In December 2009, at the Fifteenth Conference of the Parties of the UNFCCC, as part of a Copenhagen Accord involving robust GHG mitigation contributions by developed and key developing countries, the Obama administration proposed a U.S. GHG emissions reduction target in the range of 17 percent below 2005 levels by 2020 and approximately 83 percent below 2005 levels by 2050, ultimately aligned with final U.S. legislation. In January 2010, the United States inscribed its near-term proposal—to reduce emissions in the range of 17 percent from 2005 levels by 2020—in the Copenhagen Accord, formally associating itself with the Accord.

The budget information presented in this report reflects the Administration's commitment to meeting the aims of the UNFCCC while preserving a strong American economy. The President's 2011 Budget proposes over $18 billion for climate change activities. This amount is 14%

percent higher than the 2010 enacted level for climate change programs, activities, and related tax policies.

The President's Budget includes continuing support for many successful climate-related programs and initiatives as well as funding for Administration priorities that will help achieve the Administration's climate goals.

1.2. Report Outline

The President's 2011 Budget supports a wide range of climate change-related research, development, and deployment programs, voluntary partnerships, and international aid efforts. This report presents the expenditures associated with this portfolio of activities in five main categories – science, technology, international assistance, tax provisions, and adaptation efforts associated with wildlife adaptation – as described below:

- *Climate Change Science.* This category encompasses the U.S. Global Change Research Program (USGCRP), previously referred to as the U.S. Climate Change Science Program (CCSP), including funding for the Climate Change Research Initiative (CCRI).
- *Climate Change Technology.* This category comprises the U.S. Climate Change Technology Program (CCTP). The CCTP incorporates a variety of technology research, development, and deployment activities – including voluntary partnerships and grant programs – that reduce greenhouse gas emissions.
- *International Assistance.* This category describes elements of a "whole of government" approach to mobilize a wide range of resources and make use of a wide range of bilateral and multilateral assistance tools. The core budget includes resources for a coordinated set of programs designed to ensure an effective balance among the three pillars of the global climate effort-Adaptation, Clean Energy, and Sustainable Landscapes.
- *Energy Tax Provisions.* This category includes tax incentives for investments in certain energy technologies, and energy grants that can be used in lieu of certain tax credits. These incentives promote deployment of energy efficient or alternative energy technologies, which may help reduce greenhouse gas emissions.
- *Wildlife Adaptation Funding.* The Administration recognizes that there are numerous capabilities across the Federal Government to

respond to the impacts of climate change on various critical sectors, institutions and agency mission responsibilities. While the Administration is working to develop criteria to systematically account for a broader suite of adaptation programs, this interim category, described further in section 6, summarizes certain activities at the Department of the Interior designed to promote wildlife adaptation.

The following sections provide further detail in each of these five areas.

2. Climate Change Science

The U.S. Global Change Research Program (USGCRP) was mandated by Congress in the Global Change Research Act of 1990 (P.L. 101-606) to improve understanding of uncertainties in climate science, including the cumulative effects of human activities and natural processes on the environment, develop science-based resources to support policymaking and resource management, and communicate findings broadly among scientific and stakeholder communities. Thirteen departments and agencies participate in the USGCRP. The Office of Science and Technology Policy (OSTP) and the Office of Management and Budget (OMB) work closely with the USGCRP to establish research priorities and funding plans to ensure the program is aligned with the Administration's priorities and reflects agency planning. The program is currently developing a new strategic plan.

The 2011 Budget proposes $2.6 billion for the USGCRP to support the goals set forth in the program's current strategic plan. These activities can be grouped under the following areas: improve our knowledge of Earth's past and present climate variability and change; improve our understanding of natural and human forces of climate change; improve our capability to model and predict future conditions and impacts; assess the Nation's vulnerability to current and anticipated impacts of climate change; and improve the Nation's ability to respond to climate change by providing climate information and decision support tools that are useful to policymakers and the general public. Reports and general information about the USGCRP are available on the program's website, www.globalchange.gov.

Table 1. Summary of Federal Climate Change Expenditures FY 2011 Budget (Tax expenditures and budget authority, obligations, and outlays in millions of dollars)[1]

	ARRA Enacted Budget Authority	FY 2009 Actual Budget Authority	FY 2009 Obligations[2]	FY 2009 Outlays[2]	FY 2010 Enacted Budget Authority	FY 2010 Obligations[2]	FY 2010 Outlays[2]	FY 2011 Proposed Budget Authority	Proposed Change in Budget Authority 2010-2011
US Global Change Research									
Program (USGCRP)[3]	641	2,023	2,333	2,131	2,122	2,168	2,163	2,561	439
Climate Change Technology									
Program (CCTP)	25,499	5,386	15,195	4,225	5,504	15,794	13,337	5,690	186
International Assistance[4]	0	373	376	294	1,080	1,016	785	1,494	414
Energy Tax Provisions That May									
Reduce Greenhouse Gases[5,6]	N/A	1,870	1,870	1,870	4,140	4,140	4,140	3,950	-190
Energy Grants in Lieu of Tax Provisions[7,8]	N/A	1,050	1,050	1,050	3,090	3,090	3,090	4,460	1,370
Wildlife Adaptation Funding	0	0	0	0	65	47	38	87	22
Total[1,9]	26,140	10,676	20,798	9,547	15,961	26,228	23,529	18,198	2,237

[1] Budget Authority provided in millions of dollars. Discrepancies resulted from rounding and improved estimates.

[2] Obligation and outlay estimates are preliminary estimates and are subject to change, in some cases obligation and outlay estimates are based on historical obligation and outlay rates. Estimates include American Reinvestment and Recovery Act (ARRA) funding, leading to large obligations in FY 2009 and FY 2010, and large outlays in FY 2010 relative to budget authority for the same year.

[3] USGCRP was previously referred to as the Climate Change Science Program (CCSP), including funding for the Climate Change Research Initiative (CCRI).

[4] International Assistance includes activities by core agencies (i.e. Department of State, Department of the Treasury, US Agency for International Development) as well as complementary agencies (e.g. Department of Energy), but does not include Department of State, Department of Treasury, or US Agency for International Development climate assistance co-benefits nor development finance and export credit agencies.

[5] Tax incentives related to climate change included in this report are currently projected at about $15.0 billion over five years (2011-2015).

[6] Tax expenditures are estimates of the revenue losses due to a tax preference. While not exactly equivalent to budget authority, obligations or outlays, tax expenditure estimates have been included in all columns for completeness.

[7] Firms can take an energy grant in lieu of certain tax credits. The grants are considered outlays and are direct substitutes for the energy tax provisions. Estimates have been included in all columns for completeness.

[8] Energy grants in lieu of tax credits included in this report are currently projected at $11.3 billion over five years (2011- 2015)

[9] The International Assistance total contains funds that are also counted in the USGCRP and CCTP totals. Table total line excludes this double-count.

2.1. Selected Agency Highlights of the USGCRP in the 2011 Budget

- The Department of Commerce's National Oceanic and Atmospheric Administration (NOAA) is a leading sponsor of oceanic and atmospheric research and is one of the key sponsors of climate science capabilities in the Federal government. The 2011 Budget allocates \$437 million for the Department of Commerce's USGCRP efforts, predominantly from NOAA, an increase of \$77 million or 21 percent over the 2010 level.
- The National Aeronautics and Space Administration (NASA) budget includes a bold new investment in climate science. NASA's Earth Science program conducts first-of-a-kind demonstration flights of sensors in air and space in an effort to foster scientific understanding of the global integrated Earth system and to improve the ability to forecast global and regional climate change and natural disasters. The President has directed NASA to accelerate the development of new satellites that the National Research Council recommended as Earth Science priorities, in addition to flying several research satellites currently in development, conducting an airborne campaign to monitor changes in polar ice sheets, and pursuing enhancements to climate models. NASA will also develop and fly a replacement of the Orbiting Carbon Observatory (OCO), a mission designed to identify global carbon sources and sinks that was lost when its launch vehicle failed in 2009.
- The National Science Foundation (NSF) is the primary source of support for academic research for most nonbiomedical disciplines, funding basic research across the entire spectrum of the sciences and engineering. NSF USGCRP support totals \$370 million in the 2011 Budget, an increase of 16 percent over the 2010 level.
- The 2011 Budget provides \$81 million for USGCRP programs in the Department of the Interior, an increase of \$18 million or 29 percent over the 2010 funding level. Interior's lead science agency, the U.S. Geological Survey (USGS), funds several programs to monitor and analyze the impacts of climate change on natural resources, including the Global Change Effects Network and Climate Science Centers, and to develop biological and geological carbon sequestration assessments.

Table 2. U.S. Global Change Research Program
Details by Agency/Account
(Budget authority, obligations, and outlays in millions of dollars)[1]

	ARRA Enacted Budget Authority	FY 2009 Actual Budget Authority	FY 2009 Obligation[2]	FY 2009 Outlays[2]	FY 2010 Enacted Budget Authority	FY 2010 Obligation[2]	FY 2010 Outlays[2]	FY 2011 Proposed Budget Authority	Proposed Change in Budget Authority 2010-2011
Department of Agriculture									
Agricultural Research Service	0	20	20	20	20	24	20	22	2
National Institute of Food and Agriculture	0	5	5	4	56	56	50	106	50
Economic Research Service	0	0	0	0	1	1	1	1	0
Forest Service – Forest and									
Rangeland Research	0	22	22	22	32	32	32	28	-4
Subtotal – USDA[3]	0	47	47	46	109	113	103	157	48
Department of Commerce									
National Oceanic and Atmospheric									
Administration – Operations,	0	274	274	227	309	309	256	315	6
Research, and Facilities									
National Oceanic and Atmospheric									
Administration – Procurement,	218	101	234	194	49	134	111	120	71
Acquisition, and Construction									
National Institute of Standards and									
Technology (NIST)	0	2	2	2	2	2	2	2	0
Subtotal – DOC[3]	218	377	510	423	360	445	369	437	77
Department of Energy									
Science – Biological &									
Environmental Research	65	168	233	168	165	165	235	191	26
Department of Health and Human									
Services									
National Institutes of Health	0	5	5	5	4	4	4	4	0

	ARRA Enacted Budget Authority	FY 2009 Actual Budget Authority	FY 2009 Obligation[2]	FY 2009 Outlays[2]	FY 2010 Enacted Budget Authority	FY 2010 Obligation[2]	FY 2010 Outlays[2]	FY 2011 Proposed Budget Authority	Proposed Change in Budget Authority 2010-2011
Department of the Interior									
U.S. Geological Survey – Surveys, Investigations, and Research	0	45	37	43	63	20	61	81	18
Department of Transportation Federal Highway Administration –									
Federal-Aid Highways[4]	0	0	0	0	0	0	0	0	0
Federal Aviation Administration –									
Research, Engineering, and	0	2	2	1	3	3	1	3	0
Development									
Federal Transit Administration -									
Research and University Research	0	0	0	0	0	0	0	0	0
Centers[5]									
Subtotal – DOT[3]	0	2	2	1	3	3	1	3	0
Environmental Protection Agency									
Science and Technology	0	18	17	17	21	21	21	22	1
National Aeronautics and Space									
Administration									
Science	237	1,086	1,086	1,071	1,071	1,071	1,075	1,285	214
National Science Foundation									
Research and Related Activities	121	269	390	351	319	319	287	370	51
Smithsonian Institution									
Salaries and Expenses	0	6	6	6	7	7	7	11	4
U.S. Agency for International									

Table 2. (Continued)

	ARRA Enacted Budget Authority	FY 2009 Actual Budget Authority	FY 2009 Obligation[2]	FY 2009 Outlays[2]	FY 2010 Enacted Budget Authority	FY 2010 Obligation[2]	FY 2010 Outlays[2]	FY 2011 Proposed Budget Authority	Proposed Change in Budget Authority 2010-2011
Development									
Development Assistance- non-add[6]	0	17	17	17	36	36	36	43	7
Total[3]	641	2,023	2,333	2,131	2,122	2,168	2,163	2,561	439

[1] All data supersede numbers released with the 2011 President's Budget. Budget authority provided in millions of dollars. Discrepancies may result from rounding and improved estimates.

[2] All obligation and outlay data are preliminary estimates and are subject to change, in some cases obligation and outlay estimates are based on historical obligation and outlay rates.

[3] Agency subtotals and table total may not add due to rounding.

[4] The FY 2009 and FY 2011 funding for Federal Highway Administration – Federal-Aid Highways was less than $500,000.

[5] Federal Transit Administration - Research and University Research Centers is FTA's support for DOT's Center for Climate Change. The 2009 through 2011 funding amounts for this program are less than $500,000.

[6] USAID funding supports USGCRP and the Climate Change International Assistance effort. In the past, some USAID funding was counted under both categories. USAID spending is now only reported under International Assistance and does not add to the USGCRP total.

3. Climate Change Technology

The U.S. Climate Change Technology Program (CCTP) activities help to reduce, avoid, or sequester greenhouse gas emissions. The CCTP comprises research, development, and deployment efforts, including a variety of voluntary partnership and grant activities. The activities have the effect of stimulating the development and use of certain energy technologies, including renewable, low-carbon fossil, and nuclear technologies as well as energy efficient technologies, products, and process improvements.

The 2011 Budget proposes approximately $5.7 billion for the CCTP. Table 3 provides a breakdown by agency of CCTP funding.

Descriptions of some select activities are included below.

3.1. Selected Agency Highlights of the CCTP in the 2011 Budget

- *Increased Investment in DOE Climate Change Technology R&D.* The budget proposes $4.9 billion for climate change technology R&D programs at the Department of Energy, 11 percent more than the 2010 enacted level. The Department's funding supports a wide range of important R&D on key technologies such as solar, wind, nuclear and carbon capture and storage. Highlights include:
- $300 million for the Advanced Research Projects Agency – Energy (ARPA-E). ARPA-E was initiated with Recovery Act funding to pursue advances in energy technology.
- $2.355 billion for Energy Efficiency and Renewable Energy programs, an increase of $113 million over the FY 2010 enacted level. This includes increases for solar ($302 million; +$77 million or 34 percent), wind ($123 million, +$43 million or 53 percent), and geothermal energy research and development ($55 million; +$11 million or 25 percent) as well as for efforts focused on improving energy efficiency of buildings ($231 million; +$31 million or 15 percent), the development of advanced technology vehicles such as plug-in electric vehicles ($325 million; +$14 million or 4 percent), and reducing the energy intensity of industrial processes and manufacturing ($100 million; +$4 million or 4 percent).
- Establishment of new National Laboratory Carbon Capture and Storage Modeling Partnerships. The Budget provides $20 million to

advance modeling and simulation of carbon capture component technologies and geologic sequestration risk management, monitoring, verification, and accounting.

4. International Assistance

The Administration has taken a whole-of-government approach to pursue four broad international climate change financing objectives within its international assistance programs: demonstrate continued U.S. leadership in forging a global solution to the climate challenge; help developing countries focus their climate investments strategically over the coming years; create robust means of measuring, monitoring, and verifying domestic emissions in developing countries; and reduce vulnerability to climate change. Coordinating and integrating activities from across the U.S. government promotes complementarities that enhance the value of U.S. climate-related financing and increase the likelihood of successfully realizing these four objectives.

Although the Administration's efforts to address climate change are diverse, its bilateral and multilateral international climate change financing is focused on three policy pillars: adaptation, clean energy, and sustainable landscapes. These three policy pillars invest in significant emissions reduction strategies as well as activities that help communities adapt to a changing climate. Key results and indicators to measure progress have been identified for activities in these policy pillars and can be mapped to the Administration's four broad objectives. These activities will strengthen our relationships with other nations, help mitigate the security risk that climate change poses as a threat multiplier in the developing world, support our efforts for a comprehensive, multilateral approach to climate change that involves meaningful actions by all major economies, and create economic opportunities for manufacturers of clean energy technologies.

In its FY 2011 Budget, the Administration is seeking $1,384 million for core international efforts to combat global climate change, which represents a 38 percent increase over the FY 2010 level. These efforts, conducted by the U.S. Agency for International Development, the U.S. Department of State, and the U.S. Department of the Treasury, will help the most vulnerable countries respond to the growing impacts of climate change, and help forge a global solution to the climate crisis.

Table 3. Climate Change Technology Program
Details by Agency/Account
(Budget authority, obligations, and outlays in millions of dollars)[1]

	ARRA Enacted Budget Authority	FY 2009 Actual Budget Authority	FY 2009 Obligation[2]	FY 2009 Outlays[2]	FY 2010 Enacted Budget Authority	FY 2010 Obligation[2]	FY 2010 Outlays[2]	FY 2011 Proposed Budget Authority	Proposed Change in Budget Authority 2010-2011
Department of Agriculture Natural Resources Conservation									
Service – Carbon Cycle	0	9	7	7	7	7	7	7	0
National Agricultural Statistics Service	0	0	0	0	0	0	0	1	1
Agricultural Research Service –									
Bioenergy Research	0	5	5	5	5	5	5	4	-1
Cooperative State Research,									
Education and Extension Service -									
Biofuels/Biomass Research, Formula Funds, National Research Initiative	0	5	5	5	5	5	5	5	0
(National Institute of Food and									
Agriculture)									
Forest Service – Biofuels/Biomass, Forest and Rangeland Research	0	1	1	1	5	5	5	5	0
Rural Business Service – Value									
Added Producer Grants (Cooperative	0	6	2	2	4	2	2	4	0
Development Grants)									
Rural Business Service – Rural									
Energy for America Program	0	5	5	0	39	39	24	39	0
Rural Business Service – Biorefinery									
Assistance Program	0	0	0	0	0	0	0	17	17

Table 3. (Continued)

	ARRA Enacted Budget Authority	FY 2009 Actual Budget Authority	FY 2009 Obligation2	FY 2009 Outlays2	FY 2010 Enacted Budget Authority	FY 2010 Obligation2	FY 2010 Outlays2	FY 2011 Proposed Budget Authority	Proposed Change in Budget Authority 2010-2011
2008 Farm Bill, Mandatory Funding									
Rural Business Service – Rural									
Energy for America	0	55	55	19	60	0	0	70	10
National Institute of Food and									
Agriculture – Biomass Research and	0	20	20	13	28	28	20	28	0
Development									
Rural Business Service – Repowering									
Assistance Program	0	35	0	0	0	35	9	0	0
Rural Business Service – Biorefinery									
Assistance Program	0	75	0	0	245	75	4	0	-245
Rural Business Service - Bioenergy									
Program for Advanced Biofuels	0	55	0	0	55	30	8	85	30
Subtotal - mandatory funding	0	240	75	32	388	168	41	183	-205
Subtotal - discretionary funding	0	31	25	20	65	63	48	82	17
Subtotal – USDA[3]	0	271	100	52	453	231	89	265	-188
Department of Commerce									
National Institute of Standards and Technology (NIST) – Scientific and Technological Research and Services	4	7	7	7	15	15	15	15	0
NIST – Industrial Technical Services, Technology Innovation Program	0	6	6	6	1	1	1	1	0
International Trade Administration –									
Operations and Administration	0	2	2	2	2	2	2	2	0
Subtotal – Commerce[3]	4	15	15	15	18	18	18	18	0

	ARRA Enacted Budget Authority	FY 2009 Actual Budget Authority	FY 2009 Obligation2	FY 2009 Outlays2	FY 2010 Enacted Budget Authority	FY 2010 Obligation2	FY 2010 Outlays2	FY 2011 Proposed Budget Authority	Proposed Change in Budget Authority 2010-2011
Department of Defense									
Research, Development, Test and Evaluation, Army	44	98	137	124	93	98	154	28	-65
Research, Development, Test and Evaluation, Navy	18	54	69	47	13	18	29	11	-2
Research, Development, Test and									
Evaluation, Air Force	35	108	132	105	120	122	108	104	-16
Research, Development, Test and									
Evaluation, Defense Wide4	42	0	37	20	0	4	17	0	0
Subtotal – DOD[3]	139	261	376	296	226	243	308	143	-83
Department of Energy									
Energy Efficiency and Renewable	16,800	2,179	11,922	1,666	2,242	4,762	10,056	2,355	113
Energy									
Electricity Delivery and Energy Reliability	4,340	113	164	101	109	4362	595	143	34
Nuclear Energy[5]	0	787	750	677	747	745	671	793	46
Fossil Energy R&D – Efficiency and Sequestration[6]	3,400	762	598	316	560	3,810	453	545	-15
Science – Fusion, Sequestration, and Hydrogen[7]	283	700	966	818	723	740	775	738	15
Energy Transformation Acceleration									
Fund - Advance Research Projects	0	0	0	0	9	1	1	300	291
Agency- Energy (ARPA-E)[8]									
Energy Transformation Acceleration									
Fund - (ARPA-E) Recovery Act 9	400	0	3	1	0	397	54	0	0
Departmental Administration –									
Climate Change Technology Program	0	2	2	1	9	1	0	9	0
Direction[10]									

Table 3. (Continued)

	ARRA Enacted Budget Authority	FY 2009 Actual Budget Authority	FY 2009 Obligation2	FY 2009 Outlays2	FY 2010 Enacted Budget Authority	FY 2010 Obligation2	FY 2010 Outlays2	FY 2011 Proposed Budget Authority	Proposed Change in Budget Authority 2010-2011
Subtotal – DOE[3]	25,223	4,543	14,405	3,580	4,399	14,818	12,605	4,883	484
Department of Transportation National Highway Traffic Safety Administration[11]	0	0	0	0	1	0	0	0	-1
Research and Innovative Technology Administration – Research and Development	0	1	1	0	1	1	0	1	0
Federal Highway Administration – Federal Aid Highways	0	19	19	12	19	19	13	19	0
Federal Aviation Administration - RED	0	3	1	1	11	11	6	5	-6
Federal Transit Administration - Research and University Research Centers and Formula and Bus Grants[12]	100	19	19	15	94	169	20	53	-41
Subtotal – DOT[3]	100	43	41	28	125	200	39	77	-48
Environmental Protection Agency Environmental Programs and Management	0	94	97	97	113	113	113	123	10
Science and Technology	0	17	16	16	20	20	20	17	-3
Subtotal – EPA[3]	0	111	113	113	133	133	133	140	7
National Aeronautics and Space Administration Science, Aeronautics, and Exploration	31	119	119	118	124	124	122	137	13
National Science Foundation Research and Related Activities	2	24	26	23	26	26	23	27	1
Total[3]	25,499	5,386	15,195	4,225	5,504	15,794	13,337	5,690	186

[1] All data supersede numbers released with the 2011 President's Budget. Budget authority provided in millions of dollars. Discrepancies may result from rounding and improved estimates.

[2] Obligation and outlay data are preliminary estimates and are subject to change, in some cases obligation and outlay estimates are based on historical obligation and outlay rates. Estimates include American Reinvestment and Recovery Act (ARRA) funding, leading to large obligations in FY 2009 and FY 2010, and large outlays in FY 2010 relative to budget authority for the same year.

[3] Agency subtotals and table total may not add due to rounding.

[4] The American Recovery and Reinvestment Act (ARRA) provided funding to the Department of Defense for improvements in energy generation and efficiency, transmission, regulation and storage. Funding shown in 2009 includes funding that was allocated to agencies in DoD (in addition to funds provided to the military services, which is shown in other lines).

[5] Funding levels for all years shown do not include the MOX facility. The decrease in 2010 is due to the close-out of the Nuclear Power 2010 program and the termination of the Nuclear Hydrogen Initiative.

[6] The 2010 funding is for conduct of R&D only. The reductions from 2009 reflect the fact that carbon capture and storage demonstration projects are funded through ARRA.

[7] The 2009 funding includes $6 million for ARPA-E activities appropriated in the DOE Science Account. 8The 2010 funding reflects the transfer of funds originally appropriated in the science account in 2009.

[9] Funding for ARPA-E through ARRA is shown on a separate line because it has a unique account in the President's Budget Appendix.

[10] The increase from 2009 to 2010 reflects additional funding for program assessment, modeling, and analysis.

[11] The budget contains no policy recommendations for surface transportation programs subject to reauthorization, including for NHTSA and FTA programs. Instead, the budget displays baseline funding levels for all surface transportation programs.

[12] Section 3045 of the Safe, Accountable, Flexible, Efficient Transportation Equity Act: a Legacy for Users (SAFETEA-LU) establishes a National Fuel Cell Bus Technology Development Program (NFCBP). The NFCBP's goal is to facilitate the development of commercially viable fuel cell bus technologies and related infrastructure to reduce fossil fuel dependence and emissions from bus transportation. The American Recovery and Reinvestment Act, 2009, provided $100,000,000 for discretionary grants for capital investment that will assist in reducing energy consumption and greenhouse gas emissions. FTA's Transit Investments in Greenhouse Gas and Energy Reduction (TIGGER) program funds projects that demonstrate energy savings and/or GHG reductions.

The core activities are complemented by an estimated $110 million that is being sought for programs conducted by a range of additional U.S. agencies that address climate change internationally.

In addition to the funding summarized in table 4, USAID, State, and Treasury will be implementing other programs, such as food security programs or biodiversity programs, in ways that will make a significant contribution to the fight against climate change. Programs focused primarily on non-climate change goals may promote "climate-proofed" development or use adjusted techniques to significantly reduce emissions while promoting other development goals and thereby deliver climate change mitigation and adaptation co-benefits. As of April 2010, the Administration estimated the FY 2011 budget authority for these programs to be $386 million.[1] Furthermore, the Administration is enhancing U.S. efforts to address global climate change and promote clean energy technologies in important ways beyond those programs with direct appropriations. Through direct loans, loan guarantees, insurance, and working capital guarantees, U.S. development finance and export credit agencies are increasingly mobilizing investments in clean energy technologies around the world.

Based on initial estimates as of April 2010, the value of these financial products is estimated at $873 million for FY 2011.[2] These U.S. Government financial products will help American firms, financial institutions, and investors, with their foreign partners, address climate change in developing countries, offering global benefits.

Together, these activities will substantially contribute to the international community's renewed efforts to address climate change, including through the implementation of the Copenhagen Accord, and make clear the Administration's commitment to international leadership in the necessary transition to a low emission economy.

4.1. U.S. Agency for International Development (USAID)

USAID is the lead contributor to bilateral assistance, with a focus on capacity building, civil society building, governance programming and creating the legal and regulatory environments needed to address climate change. USAID will leverage its significant technical expertise to provide leadership in development and implementation of low-carbon strategies, creating policy frameworks for market-based approaches to emission reduction and energy sector reform, promoting sustainable management of agriculture lands and forests, and mainstreaming adaptation into development activities in countries most at risk. USAID has long-standing relationships with host country governments that will enable it to work together to develop shared priorities and implementation plans. USAID's engagement and expertise in agriculture, biodiversity, health, and other critical climate sensitive sectors provide an opportunity to implement innovative cross-sectoral climate change programs. Finally, USAID bilateral programs can work in key political and governance areas that multilateral agencies cannot.

4.2. Department of State

State takes the lead on diplomatic efforts and deploys financial resources in support of key multilateral and bilateral priorities. State's comparative advantage is promoting effective international solutions, advanced technology strategies, and innovative market approaches through international processes and U.S.-led diplomatic partnerships and initiatives.

4.3. Department of the Treasury

The Treasury Department is the primary vehicle by which the U.S. Government provides contributions through multilateral delivery channels, including the Climate Investment Funds and the Global Environment Facility. Multilateral assistance promotes institutional structures governed jointly by developed and developing countries, which are needed for a coordinated, global response to climate change. Multilateral institutions complement bilateral assistance by leveraging contributions from other donors, making capital investments in infrastructure, providing a range of tailored financial products, and working across a larger number of countries.

The 2011 Budget requests $400 million for the Clean Technology Fund (CTF), which aims to close the price gap in developing countries between dirtier conventional technologies and commercially available cleaner alternatives in the power sector, the transportation sector, and in energy efficiency. The CTF focus is on transforming energy use on a sector scale in the larger emitter developing countries.

In addition, the Budget requests $235 million for the Strategic Climate Fund in three programs: $90 million for the Pilot Program for Climate Resilience, $95 million for the Forest Investment Program and $50 million for

The Program for Scaling-up Renewable Energy in Low Income Countries (SREP). The Pilot Program for Climate Resilience will help finance comprehensive efforts to improve the technical capacity of countries to plan for and finance climate adaptation efforts. The Forest Investment Program will support activities informed by national plans to reduce deforestation and will focus on transitioning a small number of developing countries to participate in carbon financing for forest preservation. SREP aims to demonstrate how to put the poorest countries on a pathway that uses renewable energy to expand energy access, and stimulate economic growth.

4.4. Complementary Agencies

In addition to the core international assistance activities, a number of additional agencies provide technical and in some cases direct support for international efforts to address climate change. Two international agencies, the Millennium Challenge Corporation (MCC) and the U.S. Trade and Development Agency (USTDA), work directly with international partners on projects that may have climate change benefits.

MCC works with its foreign government partners on programs that reduce poverty through sustainable economic growth. In undertaking its poverty reduction programs, MCC seeks to integrate climate change considerations, such as adaptation and reduced emissions, where relevant. For example, MCC clean energy capital investments may support both development and emissions reductions. MCC agriculture and agricultural infrastructure programs, such as irrigation, may integrate more sustainable use of water resources at risk to become scarcer in a changing climate.

USTDA has a number of programs that couple support for U.S. exports and a focus on emissions reductions abroad. USTDA provides technical assistance to developing countries on clean energy technologies that U.S. firms provide, organizes visits for foreign entities seeking business opportunities with U.S. firms in the renewable energy sector, and funds studies on future clean energy infrastructure investments.

A number of domestic agencies with significant technical expertise complement the core international assistance activities on climate change through a variety of functions. The Department of Energy and the Environmental Protection Agency provide technical assistance on clean energy investments and environmental regulations undertaken by foreign governments; the National Oceanic and Atmospheric Administration, National

Aeronautics and Space Administration, and the National Science Foundation provide research and science assistance to the core international assistance agencies that directly supports climate change efforts; and the Forest Service works with USAID on a number of forestry programs that reduce emissions through carbon sequestration.

5. Energy Tax Provisions That May Reduce Greenhouse Gases

This report includes existing energy tax provisions and energy grants in lieu of tax provisions which may reduce greenhouse gases. Summary descriptions of the provisions are provided below and the associated revenue effects are shown in Table 5. A tax expenditure is an exception to baseline provisions of the tax structure that usually results in a reduction in the amount of tax owed. In addition to categories of tax expenditures described in previous Federal climate change expenditures reports, this report contains estimated payments from the Department of the Treasury authorized by Section 1603 of the American Recovery and Reinvestment Act.

All tax expenditure estimates presented here are based upon current tax law enacted as of December 31, 2009. Expired or repealed provisions are not listed if their revenue effects result only from taxpayer activity occurring before fiscal year 2009. The estimates reflect 2010 Budget Midsession Review economic assumptions. Legislation enacted in 2010 is not reflected in these estimates. Tax expenditure information can also be found in *Analytical Perspectives, Budget of the United States Government*, Fiscal Year 2011, Chapter 16.

New technology credit - Energy production Credit – The Tax Code provides a credit for certain electricity produced from wind energy, biomass, geothermal energy, solar energy, small irrigation power, municipal solid waste, or qualified hydropower and sold to an unrelated party.

Energy investment credit. ----- The Tax Code provides credits for investments in solar and geothermal energy property, qualified fuel cell power plants, geothermal heat pumps, small wind property and combined heat and power property. Owners of renewable power facilities that qualify for the energy production credit may instead elect to take an energy investment credit.

Table 4. International Climate Change Assistance
Details by Agency/Account
(Budget authority, obligations, and outlays in millions of dollars)[1]

	FY 2009 Actual Budget Authority	FY 2009 Obligation[2]	FY 2009 Outlays[2]	FY 2010 Enacted Budget Authority	FY 2010 Obligation[2]	FY 2010 Outlays[2]	FY 2011 Proposed Budget Authority	Proposed Change in Budget Authority 2010-2011
Core Agencies[3]								
Department of State[4]								
Diplomatic and Consular Affairs[5]	2	2	2	2	2	2	0	-2
Economic Support Fund	24	29	23	158	88	80	110	-48
International Organizations and Programs	29	29	29	39	39	39	39	0
Subtotal – State[6]	55	60	54	199	129	121	149	-50
Department of the Treasury								
Debt Restructuring – Tropical Forestry Conservation	20	20	40	20	39	39	20	0
Global Environment Facility[7]	26	26	26	26	26	26	89	63
Clean Technology Fund	0	0	0	300	300	300	400	100
Strategic Climate Fund[8]	0	0	0	75	75	75	235	160
Subtotal – Treasury[6]	46	46	66	421	440	440	744	323
U.S. Agency for International Development[4]								
Assistance for Europe, Eurasia, and Central Asia[9]	15	15	1	26	26	10	26	0
Development Assistance	113	113	94	313	313	122	446	133
Economic Support Fund	94	94	25	44	44	40	19	-25

	FY 2009 Actual Budget Authority	FY 2009 Obligation[2]	FY 2009 Outlays[2]	FY 2010 Enacted Budget Authority	FY 2010 Obligation[2]	FY 2010 Outlays[2]	FY 2011 Proposed Budget Authority	Proposed Change in Budget Authority 2010-2011
International Disaster Assistance	0	0	0	0	0	0	0	0
Subtotal – USAID[6]	222	221	120	383	383	172	491	108
Subtotal- Core Agencies[6]	323	327	240	1,003	952	733	1,384	381
Complementary Agencies[10]								
US Department of Energy[11]								
Energy Efficiency and Renewable Energy	0	0	0	8	0	0	8	0
Fossil Energy R&D – Efficiency and Sequestration	0	0	0	3	0	0	3	0
Science	0	0	0	3	0	0	3	0
Subtotal- DOE[6]	0	0	0	13	0	0	13	0
Environmental Protection Agency Environmental Programs and Management[12]	20	20	20	21	21	21	21	0
Department of Commerce								
National Oceanic and Atmospheric Administration- Operations, Research, and Facilities[13]	9	9	5	9	9	7	9	0
International Trade Administration- Operations and Administration[14]	2	2	2	2	2	2	2	0
Subtotal – Commerce[6]	11	11	7	11	11	9	11	0
National Science Foundation Research and Related Activities[15]	3	3	3	6	6	5	9	3

Table 4. (Continued)

	FY 2009 Actual Budget Authority	FY 2009 Obligation[2]	FY 2009 Outlays[2]	FY 2010 Enacted Budget Authority	FY 2010 Obligation[2]	FY 2010 Outlays[2]	FY 2011 Proposed Budget Authority	Proposed Change in Budget Authority 2010-2011
Department of Agriculture Forest Service-Forest and Rangeland Research[16]	4	4	4	5	5	5	4	-1
National Aeronautics and Space Administration Science[17]	2	2	2	2	2	2	6	4
Millennium Challenge Corporation Millennium Challenge Corporation[18]	0	0	11	2	2	5	25	23
US Trade and Development Agency Trade and Development Agency[19]	10	9	7	17	17	5	21	4
Subtotal- Complementary Agencies[6]	50	49	54	77	64	52	110	33
Total[6]	*373*	*376*	*294*	*1,080*	*1,016*	*785*	*1,494*	*414*

[1] All data supersede numbers released with the 2011 President's Budget. Budget authority provided in millions of dollars. Discrepancies may result from rounding and improved estimates.

[2] All obligation and outlay data are preliminary estimates and are subject to change. In some cases obligation and outlay estimates are based on historical obligation and outlay rates.

[3] Core agencies for the purposes of the Federal Climate Change Expenditure Report are made of the primary climate assistance activities of the Department of State, Department of the Treasury, and US Agency for International Development (USAID). The Federal Climate Change Expenditures Report contained only these core agencies in previous years.

[4] USAID and Department of State's Global Climate Change figures for 2010 and 2011 are considered estimates and may change as bureaus and missions update operational budgets and outlays are finalized.

[5] Diplomatic and Consular Affairs continues to support international climate change activities, but because it is not a foreign assistance account, it has been excluded from the international assistance crosscut, beginning in FY 2011.

[6] Agency subtotals and table total may not add due to rounding.

[7] A portion of GEF funds is allocated to programs related to climate change and shown here. The increase in FY11 is due to a new focus towards sustainable landscapes/forestry activities during the fifth GEF replenishment.

[8] The increase to the Strategic Climate Fund in FY11 reflects the budget request to increase Treasury contributions to Pilot Programs for Climate Resilience ($90m) and the Forest Investment Program, as well as a first-time contribution to Scaling-up Renewable Energy in Low Income Countries (SREP).

[9] In the 2009 Omnibus appropriation, Congress combined Assistance for Eastern Europe and the Baltic States with Assistance for the Independent States of the Former Soviet Union, making a new account called Assistance for Europe, Eurasia, and Central Asia.

[10] Estimates of complementary agencies' international climate investments are based on an initial review of funding, and may not include all relevant programs and agencies. In some cases the final review of activities, after their implementation, may change the accounting of timing and scale of financing. Additionally, development finance and export credit agencies (i.e. the Overseas Private Investment Corporation and the Export-Import Bank of the United States) make significant contributions to the deployment of clean energy technologies that are not included in this report.

[11] DOE funding represents the China and India Clean Energy Research Centers which are funded through the three accounts shown. FY10 funding reflects a reprogramming request, it was not included in the 2010 President's Budget proposal nor in 2010 appropriated funding.

[12] EPA activities include Methane to Markets, International Capacity Building, and contribution to the Multilateral Fund to support the Montreal Protocol on Substances that Deplete the Ozone Layer.

[13] NOAA activities include funding for the International Research Institute for Climate and Society (IRI).

[14] ITA funding represents activities under the Asia Pacific Partnership to promote the development and deployment of cleaner and more efficient energy technologies.

[15] NSF activities include the Regional Institutes for Global Change through the Directorate for Geosciences as well as Basic Research to Enable Agriculture Development (BREAD) through the Directorate for Biological Science.

[16] Forest Service activities include assistance to developing countries to establish and maintain sustainable landscape management.

[17] NASA activities include funding for the SERVIR initiative which consists of two web-based regional monitoring networks to provide environmental (land, sea, atmosphere, biota) information and projections to decision makers in Central America/Caribbean and East Africa.

[18] Millennium Challenge Corporation does not project future expenditures/outlays at the activity level. Its estimated outlay for FY10 is based on actual expenditures through Q1 and an initial estimate of expected expenditures through the remainder of the year.

Estimated FY10 obligations of $2 million are anticipated but subject to change. The estimated portion of MCC's total FY11 budget authority that would be used to support projects with ancillary climate benefits (estimated at $25 million) is based on an average of current, anticipated, and past climate-related activities that MCC is funding or expects to fund to date.

[19] USTDA provides funding for various forms of investment analysis and technical assistance to promote investment opportunities for U.S. companies in developing countries. USTDA has expanded its clean energy project portfolio dramatically over the last few years.

Credit and deduction for clean-fuel burning vehicles – The Tax Code allows a number of credits for certain types of vehicles and property. These are available for alternative motor vehicles (including fuel cell, advanced lean burn technology, hybrid, and alternative fuel motor vehicles), alternative fuel vehicle refueling property, and plug-ins (including plug-in electric vehicles, plug-in electric drive motor vehicles, and plug-in conversion kits).

Exclusion of utility conservation subsidies – In certain circumstances, public utilities offer rate subsidies to non-business customers who invest in energy conservation measures.

Credit for holding clean renewable energy bonds – The Tax Code provides for the issuance of Clean Renewable Energy Bonds which entitles the bond holder to a Federal income tax credit in lieu of interest. The limit on the volume issued in 2009 is $2.4 billion.

Allowance of deduction for certain energy efficient commercial building property – The Tax Code allows a deduction, per square foot, for certain energy efficient commercial buildings.

Credit for construction of new energy efficient homes – The Tax Code allows contractors a tax credit of $2,000 for the construction of a qualified new energy-efficient home that has an annual level of heating and cooling energy consumption at least 50 percent below the annual consumption of a comparable dwelling unit. The credit equals $1,000 in the case of a new manufactured home that meets a 30 percent standard.

Credit for energy efficiency improvements to existing homes – The Tax Code provides an investment tax credit for expenditures made on insulation, exterior windows, and doors that improve the energy efficiency of homes and meet certain standards. The Tax Code also provides a credit for purchases of advanced main air circulating fans, natural gas, propane, or oil furnaces or hot water boilers, and other qualified energy efficient property.

Credit for energy efficient appliances – The Tax Code provides tax credits for the manufacture of efficient dishwashers, clothes washers, and refrigerators. The size of the credit depends on the efficiency of the appliance.

Credit for residential purchases/installations of solar and fuel cells ----- The Tax Code provides a credit for the purchase of a qualified photovoltaic property and solar water heating property, as well as for fuel cell power plants, geothermal heat pumps and small wind property.

Credit for qualified energy conservation bonds ----- The Tax Code provides for the issuance of energy conservation bonds which entitle the bond holder to a Federal income tax credit in lieu of interest. The limit on the volume issued in 2009 is $3.2 billion.

Table 5. Energy Tax Provisions That May Reduce Greenhouse Gases (Revenue effect in millions of dollars)

	2009	*2010*	*2011*	*2012*	*2013*	*2014*	*2015*	*2011-2015*
New technology credit (without coal)[1]	360	780	1,020	1,240	1,530	1,530	1,500	6,820
Energy Investment Credit[2]	270	530	600	680	420	370	450	2,520
Tax credit and deduction for clean-fuel burning vehicles	130	240	260	130	170	230	390	1,180
Exclusion of utility conservation subsidies	140	140	130	120	120	120	120	610
Credit for holding clean renewable energy bonds	70	80	100	120	140	140	140	640
Allowance of deduction for certain energy efficient commercial building property	60	80	90	90	130	80	10	400
Credit for construction of new energy efficient homes	30	20	20	20	0	0	0	40
Credit for energy efficiency improvements to existing homes	570	1,950	1,460	0	0	0	0	1,460
Credit for energy efficient appliances	130	130	50	0	0	0	0	50
Credit for residential purchases/ installations of solar and fuel cells	110	180	180	180	190	190	190	930
Qualified energy conservation bonds	0	10	40	80	110	120	120	470
Tax Provisions- subtotal	*1,870*	*4,140*	*3,950*	*2,660*	*2,810*	*2,780*	*2,920*	*15,120*
Energy Grants in lieu of new technology credit or energy investment credit[3]	1,050	3,090	4,460	4,240	2,360	230	30	11,320
TOTAL (Tax provisions + grants)	*2,920*	*7,230*	*8,410*	*6,900*	*5,170*	*3,010*	*2,950*	*26,440*

[1]Estimates of revenue loss from coal provisions have been removed from the tax expenditure estimate in the budget. In previous years, the Energy Investment Credit was contained within the New Technology Credit.

[2]In previous years, the Energy Investment Credit was contained within the New Technology Credit. The Energy Investment Credit also includes the business installation of fuel cells, which was an independent entry in tables from previous years. Estimates of revenue loss from the micro-turbine provision have been removed from the tax expenditure estimate in the budget.

[3]Firms can take an energy grant in lieu of the energy production credit or the energy investment credit for facilities placed in service in 2009 and 2010 or whose construction commenced in 2009 and 2010. The grants are considered outlays and are direct substitutes for the energy tax provisions.

Energy grants in lieu of new technology credit or energy investment credit ----- Section 1603 of the American Recovery and Reinvestment Tax Act of 2009 (Section 1603) authorizes the United States Department of the Treasury (Treasury) to make payments to persons who place in service specified energy property in 2009 and 2010 or whose construction commenced in 2009 and 2010. Firms can take an energy grant in lieu of the energy production credit or the energy investment credit.

6. Wildlife Adaptation Funding

Climate change is a complex, interdisciplinary issue with the potential to affect nearly every sector and level of governmental operations. When President Obama signed the Executive Order on Federal Leadership in Environmental, Energy, and Economic Performance, on October 5, 2009, he called on the Interagency Climate Change Adaptation Task Force to develop, within one year, Federal recommendations for adapting to climate change impacts both domestically and internationally. The effort to develop recommendations is ongoing, and will inform the criteria by which climate change adaptation expenditures are estimated. While the Administration is working to develop criteria to systematically account for a broader suite of adaptation programs, this interim category summarizes certain activities at the Department of the Interior designed to promote wildlife adaptation. The 2011 Climate Change Expenditures Report includes one particular climate change adaptation activity at the Department of the Interior.

The President's Budget for FY2010 included an increase of $130 million for the Department of the Interior to fund a number of near-term activities to help fish and wildlife adapt to the impacts of climate change including planning for such changes. The FY 2011 President's Budget continues this funding including developing protocols for addressing the impacts of climate change on wildlife in land management plans and related activities.

The Department of the Interior's climate change adaptation activity is an effort to gain effective and broad collaboration to determine the causes, formulate solutions, and implement changes to reduce or reverse climate impacts to lands, waters, natural and cultural resources.

A key component of this initiative is the development of a network of Landscape Conservation Cooperatives (LCCs), which will become applied conservation science partnerships that provide scientific and technical support for spatially-explicit conservation goals and for integrated, adaptive management actions at landscape scales.

Table 6. Wildlife Adaptation Funding (Budget authority, obligations, and outlays in millions of dollars)[1]

	ARRA Enacted Budget Authority	FY 2009 Actual Budget Authority	FY 2009 Obligation[2]	FY 2009 Outlays[2]	FY 2010 Enacted Budget Authority	FY 2010 Obligation[2]	FY 2010 Outlays2	FY 2011 Proposed Budget Authority	Proposed Change in Budget Authority 2010-2011
Department of the Interior									
National Park Service - Operation of the National Park Service	0	0	0	0	10	0	0	10	0
Fish and Wildlife Service - Resource Management	0	0	0	0	40	32	32	59	19
Bureau of Land Management - Management of Lands and Resources	0	0	0	0	15	15	6	18	3
Total – Wildlife Adaptation Funding	0	0	0	0	65	47	38	87	22

[1] All data supersede numbers released with the 2011 President's Budget. Budget Authority provided in millions of dollars. Discrepancies resulted from rounding and improved estimates. Funding in the table does not include USGS climate change wildlife adaptation research, which is captured within USGCRP totals.

[2] All obligation and outlay data are preliminary estimates and are subject to change.

LCCs will engage other Federal agencies, local and State partners, and the public in crafting practical, landscape-level strategies for managing climate change impacts in coordination with the Department's Climate Science Centers. Their focus will include impacts of climate change on wildlife migration patterns, wildfire risk, drought, or invasive species that typically extend beyond the borders of any single national wildlife refuge, BLM unit, or national park, but do not encompass an area as large as an entire region.

With resident staff and through connections with partners, LCCs will develop, test, implement, and monitor conservation strategies that will be responsive to the dynamic landscape changes resulting from climate change. The LCCs will facilitate broad availability of data, modeling, and tools to land managers that will allow them to analyze and model trends in species and habitat changes. LCCs will also support improved management of water resources, historical and cultural resources, and resources that are needed by Indian Tribes and Alaska Natives.

APPENDIX

The following is a listing of Federal climate change expenditures by agency and by line item in the President's 2011 Budget Appendix. Budget Appendix line items show account level data and may not reflect sub-account level climate change information. The data in this table may be subsets of an account.

Table 7. Accounting of Federal Climate Change Expenditures by Agency
Details by Agency/Account (Tax expenditures and Budget authority, obligations, and outlays in millions of dollars)

Climate Change Expenditures by Agency[1]

	ARRA Enacted Budget Authority	FY 2009 Actual Budget Authority	FY 2009 Obligation[2]	FY 2009 Outlays[2]	FY 2010 Enacted Budget Authority	FY 2010 Obligation[2]	FY 2010 Outlays[2]	FY 2011 Proposed Budget Authority	Proposed Change in Budget Authority 2010-2011
Department of Agriculture Global Change Research Program									
Agricultural Research Service	0	20	20	20	20	24	20	22	2
National Institute of Food and Agriculture	0	5	5	4	56	56	50	106	50
Economic Research Service	0	0	0	0	1	1	1	1	0
Forest Service – Forest and Rangeland Research	0	22	22	22	32	32	32	28	-4
USDA- GCRP Subtotal[3]	0	47	47	46	109	113	103	157	48
Climate Change Technology Program									
Natural Resources Conservation Service – Carbon Cycle	0	9	7	7	7	7	7	7	0
National Agricultural Statistics Service	0	0	0	0	0	0	0	1	1
Agricultural Research Service – Bioenergy	0	5	5	5	5	5	5	4	-1

	ARRA Enacted Budget Authority	FY 2009 Actual Budget Authority	FY 2009 Obligation[2]	FY 2009 Outlays[2]	FY 2010 Enacted Budget Authority	FY 2010 Obligation[2]	FY 2010 Outlays[2]	FY 2011 Proposed Budget Authority	Proposed Change in Budget Authority 2010-2011
Research									
Cooperative State Research, Education and Extension Service - Biofuels/Biomass									
Research, Formula Funds, National	0	5	5	5	5	5	5	5	0
Research Initiative (National Institute of									
Food and Agriculture)									
Forest Service – Biofuels/Biomass, Forest and Rangeland Research	0	1	1	1	5	5	5	5	0
Rural Business Service – Value Added									
Producer Grants (Cooperative	0	6	2	2	4	2	2	4	0
Development Grants)									
Rural Business Service – Rural Energy for America Program	0	5	5	0	39	39	24	39	0
Rural Business Service – Biorefinery Assistance Program	0	0	0	0	0	0	0	17	17
2008 Farm Bill, Mandatory Funding									
Rural Business Service – Rural Energy for America	0	55	55	19	60	0	0	70	10
National Institute of Food and Agriculture									
– Biomass Research and Development	0	20	20	13	28	28	20	28	0
Rural Business Service – Repowering Assistance Program	0	35	0	0	0	35	9	0	0
Rural Business Service – Biorefinery Assistance Program	0	75	0	0	245	75	4	0	-245
Rural Business Service - Bioenergy									

Table 7. Continued

	ARRA Enacted Budget Authority	FY 2009 Actual Budget Authority	FY 2009 Obligation[2]	FY 2009 Outlays[2]	FY 2010 Enacted Budget Authority	FY 2010 Obligation[2]	FY 2010 Outlays2	FY 2011 Proposed Budget Authority	Proposed Change in Budget Authority 2010-2011
Program for Advanced Biofuels	0	55	0	0	55	30	8	85	30
Subtotal - mandatory funding	0	240	75	32	388	168	41	183	-205
Subtotal - discretionary funding	0	31	25	20	65	63	48	82	17
USDA- CCTP Subtotal[3]	0	271	100	52	453	231	89	265	-188
International Assistance									
Forest Service-Forest and Rangeland Research	*0*	*4*	*4*	*4*	*5*	*5*	*5*	*4*	*-1*
Total-USDA[3]	*0*	*322*	*151*	*102*	*567*	*349*	*197*	*426*	*-141*
Department of Commerce									
Global Change Research Program									
National Oceanic and Atmospheric									
Administration – Operations, Research, and Facilities	0	274	274	227	309	309	256	315	6
National Oceanic and Atmospheric									
Administration – Procurement,	218	101	234	194	49	134	111	120	71
Acquisition, and Construction									
National Institute of Standards and Technology (NIST)	0	2	2	2	2	2	2	2	0
DOC- GCRP Subtotal [3]	218	377	510	423	360	445	369	437	77
Climate Change Technology Program									
National Institute of Standards and									
Technology (NIST) – Scientific and	4	7	7	7	15	15	15	15	0
Technological Research and Services									
NIST – Industrial Technical Services,	0	6	6	6	1	1	1	1	0

	ARRA Enacted Budget Authority	FY 2009 Actual Budget Authority	FY 2009 Obligation[2]	FY 2009 Outlays[2]	FY 2010 Enacted Budget Authority	FY 2010 Obligation[2]	FY 2010 Outlays2	FY 2011 Proposed Budget Authority	Proposed Change in Budget Authority 2010-2011
Technology Innovation Program									
International Trade Administration –									
Operations and Administration	0	2	2	2	2	2	2	2	0
DOC- CCTP Subtotal [3]	4	15	15	15	18	18	18	18	0
International Assistance *National Oceanic and Atmospheric*									
Administration- Operations, Research, and Facilities[13]	0	9	9	5	9	9	7	9	0
International Trade Administration-									
Operations and Administration14	0	2	2	2	2	2	2	2	0
DOC- International Assistance Subtotal [3]	0	11	11	7	11	11	9	11	0
Total- Department of Commerce	222	392	525	438	378	463	387	455	77
Department of Defense									
Climate Change Technology Program									
Research, Development, Test and Evaluation, Army	44	98	137	124	93	98	154	28	-65
Research, Development, Test and Evaluation, Navy	18	54	69	47	13	18	29	11	-2
Research, Development, Test and Evaluation, Air Force	35	108	132	105	120	122	108	104	-16
Research, Development, Test and Evaluation, Defense Wide[5]	42	0	37	20	0	4	17	0	0
Total- Department of Defense	139	261	376	296	226	243	308	143	-83
Department of Energy									
Global Change Research Program									

Table 7. Continued

	ARRA Enacted Budget Authority	FY 2009 Actual Budget Authority	FY 2009 Obligation[2]	FY 2009 Outlays[2]	FY 2010 Enacted Budget Authority	FY 2010 Obligation[2]	FY 2010 Outlays[2]	FY 2011 Proposed Budget Authority	Proposed Change in Budget Authority 2010-2011
Science – Biological & Environmental Research	65	168	233	168	165	165	235	191	26
Climate Change Technology Program									
Energy Efficiency and Renewable Energy	16,800	2,179	11,922	1,666	2,242	4,762	10,056	2,355	113
Electricity Delivery and Energy Reliability	4,340	113	164	101	109	4,362	595	143	34
Nuclear Energy[6]	0	787	750	677	747	745	671	793	46
Fossil Energy R&D – Efficiency and Sequestration[7]	3,400	762	598	316	560	3,810	453	545	-15
Science – Fusion, Sequestration, and Hydrogen[8]	283	700	966	818	723	740	775	738	15
Energy Transformation Acceleration									
Fund - Advance Research Projects	0	0	0	0	9	1	1	300	291
Agency- Energy (ARPA-E)[9]									
Energy Transformation Acceleration									
Fund - (ARPA-E) Recovery Act [10]	400	0	3	1	0	397	54	0	0
Departmental Administration – Climate									
Change Technology Program Direction[11]	0	2	2	1	9	1	0	9	0
DOE- CCTP Subtotal [3]	25,223	4,543	14,405	3,580	4,399	14,818	12,605	4,883	484
International Assistance									
Energy Efficiency and Renewable Energy	0	0	0	0	8	0	0	8	0
Fossil Energy R&D – Efficiency and Sequestration	0	0	0	0	3	0	0	3	0
Science	0	0	0	0	3	0	0	3	0
DOE- International Assistance Subtotal	0	0	0	0	13	0	0	13	0

	ARRA Enacted Budget Authority	FY 2009 Actual Budget Authority	FY 2009 Obligation[2]	FY 2009 Outlays[2]	FY 2010 Enacted Budget Authority	FY 2010 Obligation[2]	FY 2010 Outlays[2]	FY 2011 Proposed Budget Authority	Proposed Change in Budget Authority 2010-2011
Total- DOE	25,288	4,711	14,638	3,748	4,564	14,983	12,840	5,074	510
Department of Health and Human Services									
Global Change Research Program									
National Institutes of Health	0	5	5	5	4	4	4	4	0
Total-HHS[3]	0	5	5	5	4	4	4	4	0
Department of the Interior									
Global Change Research Program									
U.S. Geological Survey – Surveys, Investigations, and Research	0	45	37	43	63	20	61	81	18
Wildlife Adaptation Funding									
National Park Service - Operation of the National Park Service	0	0	0	0	10	0	0	10	0
Fish and Wildlife Service - Resource Management	0	0	0	0	40	32	32	59	19
Fish and Wildlife Service - State and Tribal Wildlife Grants[2]	0	0	0	0	0	0	0	0	0
Bureau of Land Management -									
Management of Lands and Resources	0	0	0	0	15	15	6	18	3
Bureau of Indian Affairs - Operation of Indian Programs[2]	0	0	0	0	0	0	0	0	0
DOI- Wildlife Adaptation Subtotal 3	0	0	0	0	65	47	38	87	22
Total-DOI[3]	0	45	37	43	128	67	99	168	40
Department of State									
International Assistance									
Diplomatic and Consular Affairs	0	2	2	2	2	2	2	0	-2
Economic Support Fund	0	24	29	23	158	88	80	110	-48

Table 7. Continued

	ARRA Enacted Budget Authority	FY 2009 Actual Budget Authority	FY 2009 Obligation[2]	FY 2009 Outlays[2]	FY 2010 Enacted Budget Authority	FY 2010 Obligation[2]	FY 2010 Outlays[2]	FY 2011 Proposed Budget Authority	Proposed Change in Budget Authority 2010-2011
International Organizations and Programs	0	29	29	29	39	39	39	39	0
Total-State[3]	0	55	60	54	199	129	121	149	-50
Department of Transportation									
Global Change Research Program									
Federal Highway Administration – Federal-Aid Highways	0	0	0	0	0	0	0	0	0
Federal Aviation Administration –									
Research, Engineering, and Development	0	2	2	1	3	3	1	3	0
Federal Transit Administration - Research and University Research Centers	0	0	0	0	0	0	0	0	0
DOT- GCRP Subtotal [3]	0	2	2	1	3	3	1	3	0
Climate Change Technology Program									
National Highway Traffic Safety Administration	0	0	0	0	1	0	0	0	-1
Research and Innovative Technology									
Administration – Research and	0	1	1	0	1	1	0	1	0
Development									
Federal Highway Administration – Federal Aid Highways	0	19	19	12	19	19	13	19	0
Federal Aviation Administration - RED	0	3	1	1	11	11	6	5	-6
Federal Transit Administration - Research and University Research Centers and	100	19	19	15	94	169	20	53	-41
Formula and Bus Grants									
DOT- CCTP Subtotal [3]	100	43	41	28	125	200	39	77	-48
Total-DOT[3]	100	45	43	29	128	203	40	80	-48

	ARRA Enacted Budget Authority	FY 2009 Actual Budget Authority	FY 2009 Obligation[2]	FY 2009 Outlays[2]	FY 2010 Enacted Budget Authority	FY 2010 Obligation[2]	FY 2010 Outlays[2]	FY 2011 Proposed Budget Authority	Proposed Change in Budget Authority 2010-2011
Department of the Treasury									
International Assistance									
Debt Restructuring – Tropical Forestry Conservation	0	20	20	40	20	39	39	20	0
Global Environment Facility[7]	0	26	26	26	26	26	26	89	63
Clean Technology Fund	0	0	0	0	300	300	300	400	100
Strategic Climate Fund[8]	0	0	0	0	75	75	75	235	160
Total-Treasury[3]	0	46	46	66	421	440	440	744	323
Environmental Protection Agency									
Global Change Research Program									
Science and Technology	0	18	17	17	21	21	21	22	1
Climate Change Technology Program									
Environmental Programs and Management	0	94	97	97	113	113	113	123	10
Science and Technology	0	17	16	16	20	20	20	17	-3
EPA-CCTP Subtotal[3]	0	111	113	113	133	133	133	140	7
International Assistance									
Environmental Programs and Management	0	20	20	20	21	21	21	21	0
Total-EPA[3]	0	139	140	140	164	164	164	172	8
Millennium Challenge									
Corporation									
International Assistance									
Millennium Challenge Corporation	0	0	0	11	2	2	5	25	23
Total-MCC[3]	0	0	0	11	2	2	5	25	23
National Aeronautics and Space Administration									

Table 7. Continued

	ARRA Enacted Budget Authority	FY 2009 Actual Budget Authority	FY 2009 Obligation[2]	FY 2009 Outlays[2]	FY 2010 Enacted Budget Authority	FY 2010 Obligation[2]	FY 2010 Outlays[2]	FY 2011 Proposed Budget Authority	Proposed Change in Budget Authority 2010-2011
Global Change Research Program									
Science	237	1086	1086	1071	1071	1071	1075	1285	214
Climate Change Technology Program									
Science, Aeronautics, and Exploration	31	119	119	118	124	124	122	137	13
International Assistance									
Science	0	2	2	2	2	2	2	6	4
Total-NASA[3]	268	1205	1205	1189	1195	1195	1197	1422	227
National Science Foundation									
Global Change Research Program									
Research and Related Activities	121	269	390	351	319	319	287	370	51
Climate Change Technology Program									
Research and Related Activities	2	24	26	23	26	26	23	27	1
International Assistance									
Research and Related Activities	0	3	3	3	6	6	5	9	3
Total-NSF3	123	293	416	374	348	348	312	403	55
Smithsonian Institution Global Change Research Program									
Salaries and Expenses	0	6	6	6	7	7	7	11	4
Total- Smithsonian3	0	6	6	6	7	7	7	11	4
US *Trade and Development*									
Agency									
International Assistance									
Trade and Development Agency	0	10	9	7	17	17	5	21	4

	ARRA Enacted Budget Authority	FY 2009 Actual Budget Authority	FY 2009 Obligation[2]	FY 2009 Outlays[2]	FY 2010 Enacted Budget Authority	FY 2010 Obligation[2]	FY 2010 Outlays[2]	FY 2011 Proposed Budget Authority	Proposed Change in Budget Authority 2010-2011
Total-TDA[3]	0	10	9	7	17	17	5	21	4
U.S. Agency for International									
Development									
Global Change Research Program									
Development Assistance	0	17	17	17	36	36	36	43	7
International Assistance									
Assistance for Europe, Eurasia, and Central Asia[9]	0	15	15	1	26	26	10	26	0
Development Assistance	0	113	113	94	313	313	122	446	133
Economic Support Fund	0	94	94	25	44	44	40	19	-25
International Disaster Assistance	0	0	0	0	0	0	0	0	0
USAID-International Assistance Subtotal3	0	222	221	120	383	383	172	491	108
Total- USAID3	0	222	221	120	383	383	172	491	108
Total All Agencies	26,140	7,756	17,878	6,627	8,731	18,998	16,299	9,788	1,057
Energy Tax Provisions That May Reduce									
Greenhouse Gases	N/A	1,870	1,870	1,870	4,140	4,140	4,140	3,950	-190
Energy Grants in lieu of new technology credit or energy investment credit	N/A	1,050	1,050	1,050	3,090	3,090	3,090	4,460	1370
Total All Agencies + Tax Provisions	26,140	10,676	20,798	9,547	15,961	26,228	23,529	18,198	2,237

[1]All data supersede numbers released with the 2011 President's Budget. Budget authority provided in millions of dollars. Any discrepancies are the result of rounding and improved estimates.

[2]All obligation and outlay data are preliminary estimates and are subject to change, in some cases obligation and outlay estimates are based on historical obligation and outlay rates.

[3]Agency subtotals and table total may not add due to rounding and subtraction of double-counts.

End Notes

[1] Summary of the "whole of government" U.S. International Climate Change Financing is available at http://www.state.gov/documents/organization/140689.pdf

[2] Estimates of development finance and export credit agencies' international climate investments are based on an initial review of planned projects, and in some cases the final review of activities, after their implementation, may change the accounting of timing and scale of financing.

INDEX

D

E

F

G

H

I